PROCESS INTENSIFICATION FOR THE CHEMICAL INDUSTRY

This volume consists of papers presented at the 1st International Conference on Process Intensification for the Chemical Industry, Practical Applications of New Technology, held in Antwerp, Belgium, on 6-8 December 1995 organized by BHR Group Limited.

TECHNICAL ADVISORY COMMITTEE

COSPONSORING ORGANIZATIONS

CEFIC, the European Chemical Industry Council
SUSTECH, Collaborative Research and Development in Sustainable Technologies for the Process Industries

ORGANIZED AND SPONSORED BY

BHR Group Limited
Cranfield
Bedford
MK43 0AJ
UK

Tel: 01234 750422
Fax: 01234 750074

1ST INTERNATIONAL CONFERENCE ON

Process Intensification for the Chemical Industry

Edited by C Ramshaw

BHR Group Conference Series
Publication No.18

Papers presented at the *1st International Conference on Process Intensification for the Chemical Industry*, organized and sponsored by BHR Group Limited, and held in Antwerp, Belgium on 6–8 December 1995.

Mechanical Engineering Publications Limited
LONDON

ISBN 0 85298 978 4

Printed and bound in Great Britain by Antony Rowe Ltd, Chippenham

A CIP catalogue record for this book is available from the British Library.

Other titles in the BHR Group Conference Series:

1 *Offshore Loss Prevention – A Systematic Approach*
Edited by C P A Thompson
2 *Fluid Power The Future for Hydraulics*
Edited by N Way
3 *Effective Membrane Processes – New Perspectives*
Edited by R Paterson
4 *Multiphase Production*
Edited by A Wilson
5 *Bioreactor and Bioprocess Fluid Dynamics*
Edited by A W Nienow
6 *Slurry Handling and Pipeline Transport – Hydrotransport 12*
Edited by C A Shook
7 *Pipe Protection*
Edited by A Wilson
8 *Advances in Water and Effluent Treatment*
Edited by M J D White
9 *Fluid Sealing*
Edited by B Halligan
10 *Water Pipeline Systems*
Edited by D S Miller
11 *Hydraulic Modelling*
Edited by A J Saul
12 *Aerodynamics and Ventilation of Vehicle Tunnels*
Edited by I J Cockram
13 *Jet Cutting Technology – Applications and Opportunities*
Edited by N G Allen
14 *Multiphase 95 - Where are We on the S Curve?*
Edited by A Wilson
15 *Loss Prevention in the Oil and Gas Industry*
Edited by C Thompson
16 *Pipeline Protection*
Edited by A Wilson
17 *Water and Waste Water Treatment*
Edited by M White

CONTENTS

1st International Conference on
PROCESS INTENSIFICATION FOR THE CHEMICAL INDUSTRY.
Antwerp, Belgium: 6–8 December 1995

INTRODUCTION

The Incentive for Process Intensification

Process Intensification (PI) is a term used to describe the strategy of making dramatic reductions in the size of a chemical plant in order to reach a given production objective. The concept was pioneered in ICI during the late 70s, when the primary goal was to reduce the capital cost of a production system. The virtue of the PI approach will be recognised when it is appreciated that the Main Plant Items involved in the process (ie reactors, heat exchangers, separators etc) only contribute around 20% of the cost of a given plant. The balance is incurred by installation costs which involve pipework, structural support, civil engineering and so on.

A major reduction of equipment size, coupled preferably with a high degree of "telescoping" of equipment function - for example reactor/heat exchangers or combined condenser/distillation/reboilers - could generate very significant cost savings by eliminating support structure, expensive column foundations and long pipe runs.

The degree of miniaturisation involved is that needed to generate the cash savings required. Volume reductions of the order of >100 must be our target in order to secure the desired impact on costs. While an individual "intensified" unit may cost a little more than the conventional equivalent (though hopefully it will not) it <u>must</u> generate substantial overall savings in the cost of the process system. In addition, the PI philosophy should be applied across the whole spectrum of unit operations used in the plant. Bearing in mind the dramatic size reduction which is sought, PI will probably involve novel and unusual approaches to equipment design. It is not a strategy for faint-hearted apple polishers! Herein lies one of its main disadvantages, notably the lack of new design codes to engender confidence in those who specify new equipment. Radical and unconventional approaches will be the order of the day. Many searching questions will be posed such as the need for turbulent flow in pipes, the use of batch rather than continuous operation in "desk top" plant and the application of merely terrestrial acceleration to multiphase systems, to name but a few.

It is a sobering thought that if chemical engineers were given a free hand to design the human digestive and elimination system, our bodies would be very large and require many kilowatts to operate them. On the other hand nature operates unobtrusively with laminar flow in high density matrices (kidneys/lungs) on a semi continuous basis, and copes with fouling problems by coughing! As scientists and engineers, we should not be too arrogant to learn a few lessons from the natural world.

While cost reduction was the original target for PI, it quickly became apparent that there were other important benefits, particularly in respect of improved intrinsic safety, reduced environmental impact and energy consumption. Given the anticipated plant volume reductions, the toxic and flammable inventories of intensified plant are correspondingly reduced, thereby making a major contribution to intrinsic safety. As has been said by T Kletz, "What you don't have you can't leak".

With regard to the environment, the intensified plant of the future will be much less obtrusive, with the distillation/absorption towers of our present chemical complexes being replaced by more compact and inconspicuous equipment which may be hidden by the boundary tree line. In addition, the cost of effluent treatment systems will be less, allowing tighter emission standards to be reached economically. However, the most telling environmental influence of PI could well be in the development of new reactor designs for truly clean technology. Rather than accept mere "end of pipe" solutions, we must create fluid dynamic environments which allow the intrinsic chemical kinetics free rein. We then have a far better prospect of designing reactors which operate intensively and which give high selectivity. This would facilitate delivery of a high quality product without an expensive downstream purification sequence. PI has further benefits and implications for the overall business process of chemical production. Thus the very short residence times involved (seconds rather than hours) will inevitably result in a very rapid response to altered process conditions. The ability to follow market demand is therefore improved. In addition, rather than transport hazardous chemicals, it may be feasible to manufacture them economically on a customer's site, as is currently done for tonnage oxygen and nitrogen.

Having extolled the potential benefits of PI, it only remains for the R&D community in academia and industry to deliver the goods! While the papers presented at this conference show that the challenge has been accepted enthusiastically, we need the support and commitment of the relevant national and European funding agencies which share the PI vision. It is significant that in China the Ministry for Chemical Industry is basing its next five year plan on PI, with particular emphasis on "Higrau" (ie Higee). A research institute has been established at the Beijing University of Chemical Technology to develop the relevant technology.

Arguably one of the more impressive historical examples of PI was George Stephenson's development of the mobile steam engine. His two critical contributions were the use of a multi tube boiler and the steam blast ejector for the flue gases. These innovations were directly responsible for the triumph of his prototype locomotive and the subsequent arrival of the railway era in the nineteenth century. Perhaps PI could exert a similar influence on the twenty first century.

Professor C Ramshaw, University of Newcastle upon Tyne, UK
September 1995

HEAT

TRANSFER

1st International Conference on:
Process Intensification for the Chemical Industry
Antwerp, 6-8 Dec. 1995

Application of Laminar Flow Heat Transfer Enhancement

G.T.Polley & I.Gibbard
Cal Gavin Ltd.
Birmingham

There are three common misconceptions regarding the application of laminar flow heat transfer enhancement. Firstly, it is thought that it is only worthwhile when high viscosity fluids are being handled. Secondly, it is commonly stated that it must be paid for by additional pressure drop. Thirdly, it is assumed that heat transfer enhancement is only worthwhile when applied a stream having a controlling heat transfer coefficient. In this paper we attempt to dispell all of these views.

Mechanical constraints play a significant role in the design of shell-and-tube heat exchangers. For instance, it is normal practice in exchanger design to restrict the length of the tubes to less than 6 metres. This mechanical limitation has at least two effects on exchanger design. First, it means that with plain tube heat exchangers, the amount of pressure drop that can be used to promote heat transfer is severely limited. It is not uncommon to find that the largest part of the actual tubeside pressure drop is that associated with flow through the exchanger nozzles. Secondly, duties involving large surface areas subsequently lead to large tube counts. If such tube counts lead to low Reynolds Numbers the designer then increases the number of tube side passes in order to maintain turbulent flow within the tubes; the theory being that turbulent flow is necessary for reasonable heat transfer.

The use of multiple tube passes has four detrimental effects. First, it leads to a reduction in the number of tubes that can be accommodated in the shell. Second, for bundles having more than four tube passes, the pass partition lanes introduced into the bundle give rise to increased bundle bypassing and a reduction in the shellside heat transfer coefficient. Thirdly, it gives rise to wasted use of tubeside pressure drop in the return headers. Finally, multiple tubepasses introduces the need to examine temperature cross considerations. Here again it is often found that the duty has to be split between a number of shells in series.

Now the premise has been that turbulent flow is needed for good tubeside heat transfer. This is not the case. The heat transfer coefficient achieved at a Reynolds Number of 2000 using a HiTRAN wire matrix insert to process a light organic liquid (viscosity of 0.55 cps, Prandtl Number of 5.9) is 1 945 W/sq.m. K. A Reynolds Number of 28 000 would be needed to achieve this coefficient in a plain tube.

The pressure drop incurred in the tube fitted with HiTRAN would be 3.5 kPa/m. That incurred in the plain tube operating at the higher velocity would be 3.98 kPa/m.

The pressure loss incurred by the enhanced tube is lower than that of the plain tube even at fixed length. If the higher plain tube velocity has been bought at the expense of increased tube passes, it is seen that the plain tube pressure drop is over ten times greater than the enhanced tube pressure drop for the same performance.

These advantages translate into design. Table 1 presents a comparison of three designs recently developed for lean/rich exchangers on an aromatics plant. The first design is a plain tubed unit. The second is a HiTRAN unit handling the duty in the minimum number of shells. The third is a HiTRAN unit handling the duty with the minimum heat transfer area. Despite the additional mechanical design complexity involved with moving to a single tube pass, the HiTRAN units show signficant cost savings.

It is clear that laminar flow enhancement has a role to play in the processing of low viscosity fluids and the pursuit of turbulence to promote heat transfer can be very wasteful.

Evaluating Heat Transfer Enhancement

The misconception regarding pressure drop penalties associated with the use of heat transfer enhancement has arisen because at the same Reynolds Number the enhanced tubes invariably exhibit higher friction factors found with than a plain tube. However, as seen above, enhanced tubes are generally operated at lower Reynolds Numbers than plain tubes. The result is higher heat transfer coefficients for a given pressure drop.

Different enhancement devices will operate at different Reynolds Numbers. These differences in operational Reynolds Numbers mean that the comparative performance of differing devices cannot be directly determined by an examination of heat transfer j-factor and friction factor plots. The application of a simple algorithm is necessary.

The heat transfer performance of many (if not most) enhancement devices can be correlated over quite large Reynolds Number ranges by an equation of the form:

$$j_H = a\,Re^{b} \qquad (1)$$

and, friction factors by an equation of the form:

$$f = x\,Re^{-y} \qquad (2)$$

As shown previously [1], given equations of this form, heat transfer coefficient, pressure drop and heat exchanger area are related by the equation:

$$\Delta P = \frac{K}{M} A\,h^{m} \qquad (3)$$

where:

$$m = \frac{3-y}{b} \qquad (4)$$

A full derivation of this equation is presented in the Appendix.

Table 1.

	Plain Tube Design	HiTRAN Minimum Shells	HiTRAN Minimum Area
TEMA Type	AEU	AEL	AEL
No. Shells	3	1	2
Area (sq.m.)	1092	742	560
Tube length (m)	6	6	6
Tubeside Press. Drop (kPa)	138	16	138
Shellside Press. Drop (kPa)	94	25	95

For assumed shellside heat transfer and fouling resistances the performance relationships for the various enhanced tubes can be compared with both each other and that of the plain tube.

A convenient means of making these comparisons is to plot exchanger area against an exchanger 'duty factor'. The duty factor is a determined by multiplying the tube pressure drop by the ratio of the exchanger's effective mean temperature difference to the absolute value of the tubeside temperature change. (The effective mean temperature difference is the product of the log mean temperature difference and the muti-pass correction factor).

The pressure drop relationship (equation 3) can be combined with the basic heat exchanger design equation (which relates heat duty, LMTD, overall heat transfer coefficient and area) to give:

$$\frac{\Delta P . \Delta T_{LM}}{\Delta t} = k c_p \left(\frac{1}{h} + R \right) h^{m} \qquad (5)$$

This equation shows that for a given duty factor, the tubeside heat transfer coefficient is independent of mass flowrate. It dependent on the physical properties of the fluid and the 'opposing thermal resistance'. The opposing thermal resistance consists of the sum of the other film resistances present in the exchanger (fouling resistances, wall resistance and film resistance on the opposite side of the exchanger).

The lower the duty factor the larger is the area required to treat each unit mass flow of fluid.

Fixing the duty factor also fixes the exchanger tube length. For, if the duty factor fixes the heat transfer coefficient it also fixes the velocity. Fixed velocity and fixed pressure drop in turn require a fixed tube length.

The relationship between heat transfer area and throughput at fixed duty factor is linear. The duty factor fixes the overall heat transfer coefficient. The effective mean temperature difference is fixed by the process. The heat load is a linear function of throughput.
exists a linear.

A plot of area per unit mass flow against duty factor is therefore a very convenient means of comparing exchangers having plain tubes with those employing heat transfer enhancement devices.

The dependance of the area on stream physical properties and the opposing resistance means that each exchanger application has its own individual 'area/duty factor' plot. However, general guidance can be provided through the developments of plots for specific fluids.

In this work we present plots for heat transfer to water for three different opposing resistance: a case in which both sides of the unit have resistanes of similar magnitude (R=0.001), a case in which the tubeside resistance is the controlling resistance (R=0.0002), and a case in which the shellside resistance is the controlling resistance (R=0.01).

Case 1. Both Resistances Significant

The size/duty factor plot for this situation is presented in Figure 2. Four types of unit are considered: a plain tubed unit, one using twisted tape inserts, one using a wire matrix insert, and one using a static mixer. There is very little difference between the performance of the plain tube and one fitted with a twisted tape insert (note: these comparisons are on the basis of pressure drop not Reynolds Number). The tubes fitted with laminar flow enhancement devices both lead to lower area needs.

In Figure 2 the bounds set by performance and mechanical design considerations are shown.

The geometrical limit has been set at a tube length of 48m. This is equivalent to an eight tube pass unit of maximum tube length.

The wire matrix inserts operate well up to a Reynolds Number of 5 000. Due to a lack of data, the application of static mixers has been considered to be limited to Reynolds Numbers below 2200.

The scope for using twisted tapes now becomes more clear. The plain tube can only absorb pressure drop equivalent to a duty factor of 1.0 in a single shell. The twisted tape insert allows pressure drop to be absorbed up to a duty factor of 5.2.

Using two shells in series, pressure drop up to a duty factor of 10.4 can be used to promote heat transfer in plain tubes.

But what about the laminar flow devices? The static mixer reaches its limiting Reynolds Number at a duty factor of 0.8. At this point the required exchanger area is 25 sq.m. for every kg/s of flow. This compares with the area need of a plain tube at its single shell limit of 33.4 sq.m. for every kg/s of tubeside material processed.

The area needed for a two shell, plain tube unit operating at its limit is 27 sq.m. s/kg.

The HiTRAN insert is still well able to absorb pressure drop at a duty factor of 12. At this point it is operating at a Reynolds Number of 3775 and requires 23.8 sq.m. of surface for each kg/s of material processed. This is equivalent to a 29% reduction on the single shell plain tube unit.

Case 2. Tubeside Controlled

It is generally asumed that tubeside heat transfer enhancement is best suitable for situations in which the tubeside heat transfer coefficient is controlling. This was investigated by repeating the above exercise using an opposing resistance of 0.0002. The results are presented in Figure 3.

In this situation the plain tubed unit is able to absorb pressure drop up to a duty factor of 12 (at which the area requirement is 8.34 sq.m. s/kg). The HiTRAN insert reaches a Reynolds

Number of 5000 at a duty factor of 6.8 (at which it has an area requirement of 6.5 sq.m. s/kg).

There is indeed benefit from using heat transfer enhancement, even with a substance such as water which generally has provides for good heat transfer. However, the benefits are probably less than found in the previous case!

Case 3. Shellside Controlled

It is generally assumed that heat transfer has no role to play when the shellside heat transfer coefficient is controlling. To investigate this viewpoint the exercise was repeated using an opposing resistance of 0.01 sq.m. K/W.

The results are shown in Figure 4. With plain tubes it was found that at the lower duty factors it was not possible to achieve turbulent flow. At a duty factor of 2.0 kPa the plain tube Reynolds Number was 2617. However, the exchanger would need three exchanger shells in series to achieve the required duty.

Despite the fact that the unit has a low shellside coefficient, tubeside heat transfer enhancement is necessary to get a reasonable tubeside performance.

Discussion

The above results apply to heat transfer to and from water flowing through shell-and-tube exchanger tubes. They are confirmed by design checks undertaken using a state-of-the-art commercial heat exchanger design program.

Many engineers would not consider heat transfer enhancement to be of potential benefit in this situation. The results indicate that in many design cases they would be mistaken.

The application of heat transfer enhancement has greater significance than simply increasing a heat transfer coefficient. Its use must also be considered in the context of mechanical design complexity. Laminar flow heat transfer devices have the following benefits:

1. They lead to size reduction;
2. They allow the designer to reduce the number of tubeside passes required for a given duty;
3. They allow the designer to reduce the number of shells in series required for a given duty;
4. They allow the designer to make fuller use of available pressure drop (pressure drop that would otherwise be wasted by passage through valves etc.).

The case for using heat transfer enhancement with water is strong. The case for using it with more viscous fluids is overwhelming.

Reference

Polley G.T., Nasr J. & Terranova A. 'Determination and Application of the Benefits of Heat Transfer Enhancement', Trans.I.Chem.E. 1994,72A,616-620

Figure 1

Duty Factor/Area Plot (R=0.001)

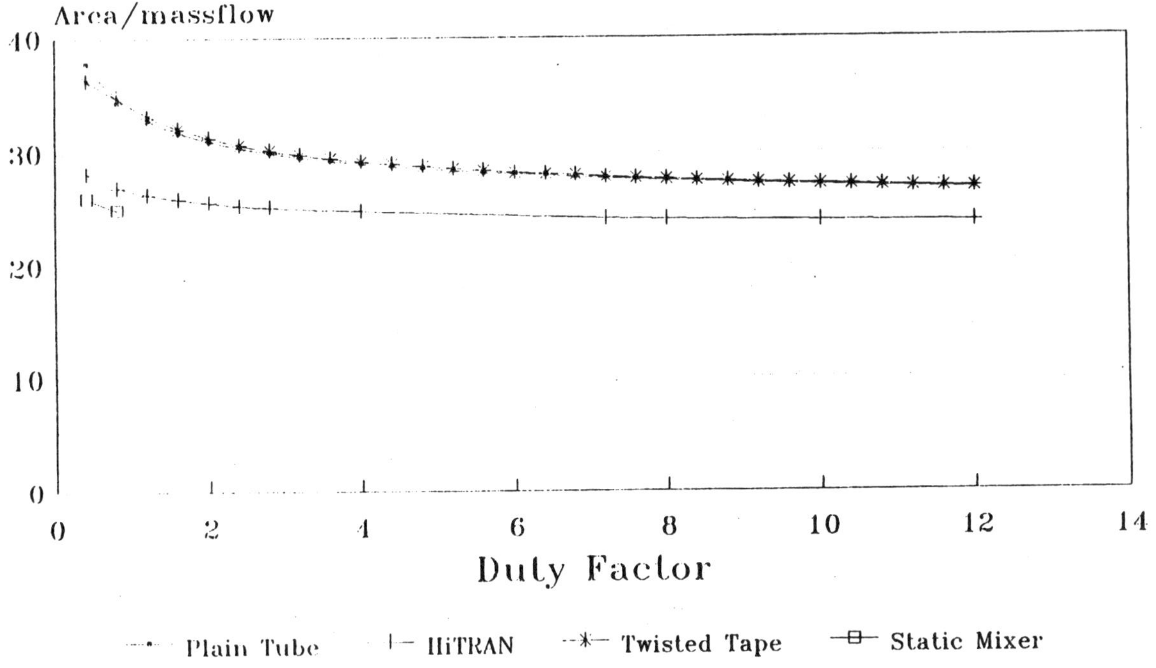

Water

Figure 2

Duty Factor/Area Plot (R = 0.001)

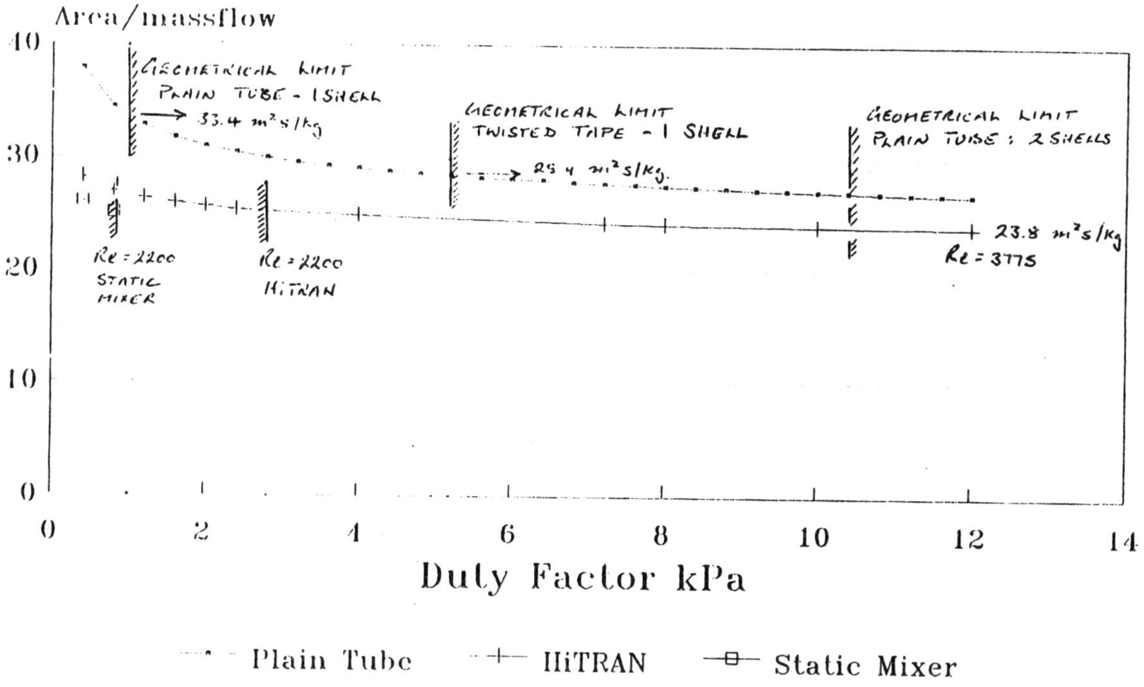

Water

Figure 3

Duty Factor/Area Plot (R = 0.0002)

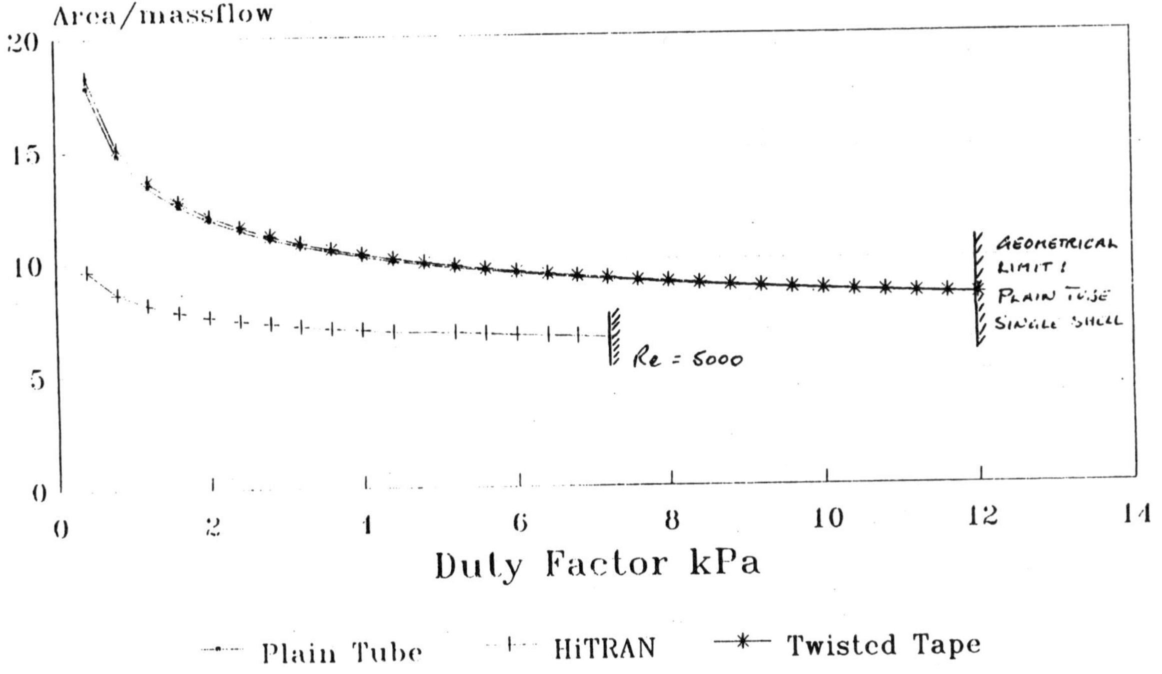

Water

Figure 4

Duty Factor/Area Plot (R=0.01)

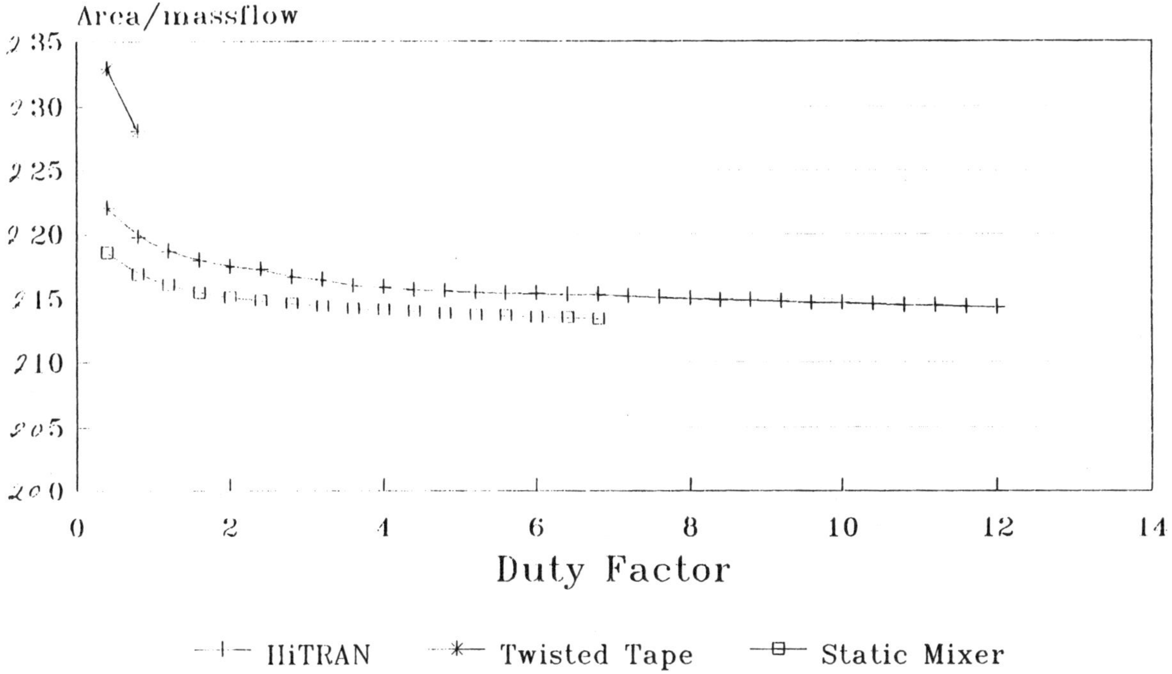

Water

Appendix

Pressure Drop/Area Relationship

The pressure drop associated with the flow of a fluid through an exchanger tube is given by:

$$\Delta P = 2f\{\frac{l}{d}\}\rho u^2 \qquad (1)$$

where, f, is the friction factor. The equation applies to both plain tubes and to tubes fitted with heat transfer enhancement devices. However, each arrangement has its own friction factor relationship.

The friction factor can be correlated against Reynolds Number, across quite large Reynolds Number ranges, by a simply power law expression:

$$f = xRe^{-y} \qquad (2)$$

Combination of these two equations yields the relationship between pressure drop and velocity:

$$\Delta P = k_1 l u^{2-y} \qquad (3)$$

where:

$$k_1 = 2x\{\frac{\rho}{d}\}\{\frac{\mu}{\rho d}\}^y \qquad (4)$$

Heat transfer coefficient can also be related to velocity. Again, across large Reynolds Number ranges, the heat transfer j-factor is related to Reynolds NUmber by a power law:

$$j = aRe^b \qquad (5)$$

Extracting the film coefficient from this equation yields:

$$h = k_2 u^b \qquad (6)$$

where:

$$k_2 = a\{\frac{\lambda}{d}\} Pr^{1/3}\{\frac{\rho d}{\mu}\}^b \qquad (7)$$

Heat transfer surface is given by:

$$A = N_t \pi d l \tag{8}$$

The number of tubes can be related to mass throughput and velocity:

$$N_t = \{\frac{M}{\rho}\}\{\frac{4}{\pi d^2}\}u^{-1} \tag{9}$$

So, by combining equations 8 and 9 it is possible to relate tube length with velocity:

$$l = \{\frac{k_3}{M}\}Au \tag{10}$$

where:

$$k_3 = \frac{\rho d}{4} \tag{11}$$

This equation for length can be introduced into equation 3 to yield:

$$\Delta P = k_1 k_3 \{\frac{A}{M}\}u^{3-y} \tag{12}$$

Velocity and heat transfer coefficient have been related in equation 6, so:

$$\Delta P = \{\frac{k_1 k_2}{k_3^m}\}\{\frac{A}{M}\}h^m \tag{13}$$

where:

$$m = \frac{3-y}{b} \tag{14}$$

PROCESS INTENSIFICATION: SPIRAL POLYMER FILM COMPACT HEAT EXCHANGER

R.J. Jachuck & C. Ramshaw
Department of Chemical & Process Engineering
University of Newcastle upon Tyne
Newcastle upon Tyne, NE1 7RU, UK

Abstract: This paper is concerned with the design, development and performance study of a novel spiral polymer film compact heat exchanger (SPFCHE) for air/water duties. Corrugated Poly-ether ether ketone (PEEK) films of thickness 100µm have been used and the performance of the heat exchanger studied by varying the air flow rate from 35 l/min to 690 l/min, and the water flow rate from 200 l/min to 650 l/min. Design data in the form of J_h and f factors have also been presented.

1. Introduction

The concept of all - polymer heat exchangers is still in its developmental stage(1). At Newcastle PEEK films which have a working temperature range up to 250 °C and have good chemical resistance coupled with intrinsically superior fouling resistance, have been used to develop a cross flow cross corrugated compact heat exchanger(2). The results of the above studies have suggested that polymer films could be extensively applied for many heat transfer applications involving modest temperatures. The study also suggested that the production of cost effective PFCHE's would involve the innovative application of fabrication and bonding techniques used in industry today. Although the heat transfer achieved for a given pressure drop is competitive compared with that of more conventional exchangers it must still be recognised that the pressure/drop unit flow path length is comparatively high. This suggests that in order to keep the pressure drop to a minimum the design of such heat exchangers must incorporate a short path length but a large approach area. The success of the first prototype at Newcastle University has fuelled interest by other parties in application of PFCHE's for domestic heating and ventilation duties, use in automobiles and in process industries where the environment is chemically aggressive.

For most domestic applications and for car radiators a high pressure drop on the air side renders the design impractical and uneconomical. It therefore important to design an exchanger which has very short flow path length for the air side whilst providing high surface area densities.

2. Description of the Test Facility

As shown in Figure 1. two long, narrow strips of 100μm thick PEEK films were corrugated, with an amplitude of 2mm and corrugation height of 1mm, and bonded together around the edges to form a channel for the fluid to flow through. In the test rig this bonding was done with silicoset but with the appropriate equipment thermal bonding would be possible and would give the rig a much improved strength. The corrugations would ideally be at 90° to each other but as this would render the strip unable to bend, therefore a compromise between fabrication and heat transfer was made and the channels form a chevron pattern at 20°.

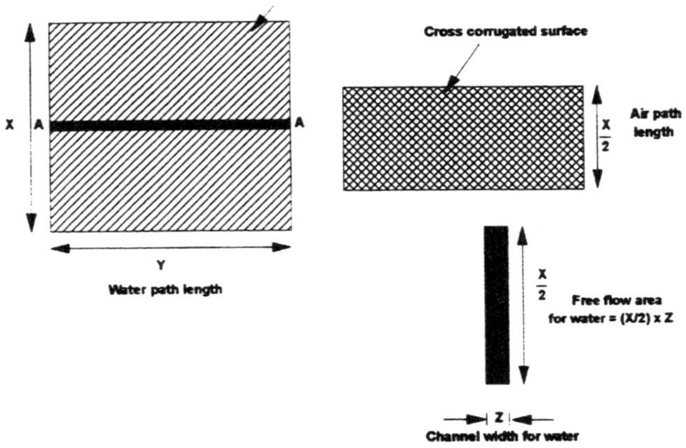

Figure 1. Design of Spiral Strip.

As well as bonding the strips along the edges, spacers were inserted to keep the sheets separated in the middle thus reducing water side pressure drop, as shown in Figure 2. The spacers also have the advantage of stopping the water channel from 'bowing' when large water flow rates are employed. The spacers were 1 cubic mm blobs of silicoset placed at 100μm alternate intervals down each edge. It was also thought that these spacers could act as baffles and promote uniform distribution and good mixing characteristics within the water stream.

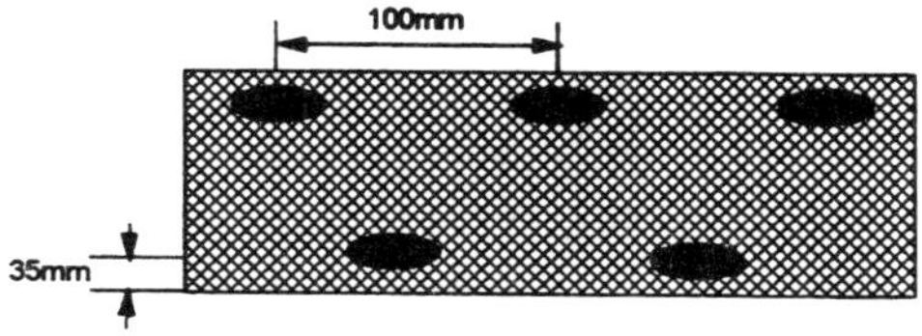

Figure 2. Layout of the water side spacer.

Figure 3 shows the final assembly of the heat exchanger with its end caps. The strips of PEEK were spiralled around a mandrel which formed the air path through the spaces between the alternate layers. The air path length was very short(10cm) and therefore resulted in low pressure drops on the air side. The spiral was held together with end caps into which the air line was directed. The water flow was through inlet and outlet ports.

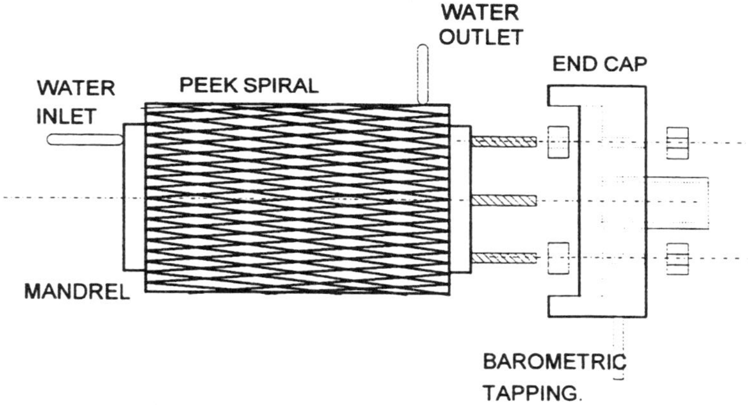

Figure 3. Sectional view of the heat exchanger

The water inlet was from the main supply via a rotameter and stopcock and was discharged to the drain. As shown in Figure 4 the air flow was in total cross flow with the water. Physical details of the heat exchanger can be seen in Table 1.

Table 1. Details of the Spiral Heat Exchanger

Air Flow Path Length.	0.12 m
Water Flow Path Length.	0.98 m
Superficial Heat Transfer Area.	0.093 m^2
Hydraulic Diameter.(Dh))	0.66 mm
Corrugation Wavelength.	0.002 m
Corrugation Amplitude.	0.0005 m
Facial Area for Air Flow.	7.15 E-05 m^2
Facial Area For Water Flow.	1.0 E-04 m^2

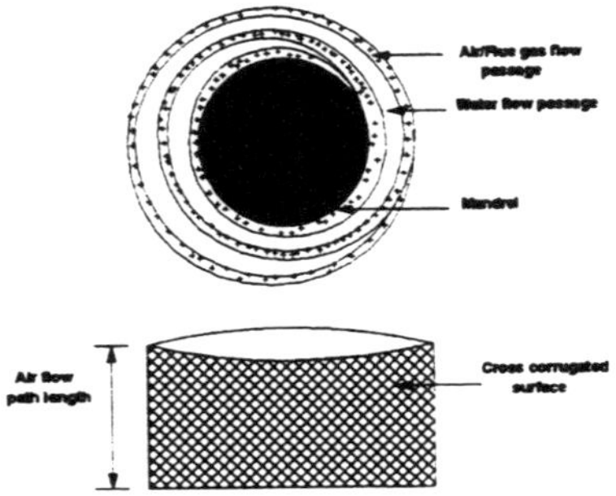

Figure 4. Mounting Arrangement for the Spiral Matrix.

3. Results and Discussion

Experimental data were collected at constant air flow rates for a range of water flow rates, and also at constant water flow rate for a range of air flow rates. The variables measured were inlet and outlet water and air temperatures, and pressure drop measurements across the heat exchanger. From the data acquired values for the overall heat transfer coefficients (U) were calculated over a range of Reynolds numbers for both the air and the water streams. Using these U values, film heat transfer coefficients were calculated for both sides and the j_h factor ($j_h = St^*Pr^{2/3}$) evaluated for corresponding Reynolds numbers. Using the data obtained from experimental runs design correlations for the j_h and f factors were calculated and have been have presented in Tables 2 and 3 respectively.

Table 2 : Heat Transfer Factor, j_h, Correlations

Flow Side	Correlations	Reynolds Number Range
water side	$j_h = 0.2145 * Re^{-0.886}$	10 - 80
air side	$j_h = 0.0897 * Re^{-0.275}$	100 - 1000

Table 3 : Friction Factor Correlations

Flow Side	Correlation
Air Side	$0.299 * Re^{-0.575}$
Water Side	$200.65*Re^{-1.266}$

Having obtained the heat and the pressure drop characteristics of the spiral heat exchanger a case study was carried out to investigate the possibility of using spiral polymer heat exchangers as an alternative car radiator. The current favoured design for a car radiator is a metallic flat tube-and-fin heat exchanger which is the most compact and efficient of those available to the motor industry. The restriction on the radiator is one of size and it is usually found at the very front of the engine. This is because the air is drawn in clean from the atmosphere rather than having passed through the engine first, thus not depositing any oil-fouling on the finned surface of the radiator. The temperature of the fluid through a combustion engine must be kept at a fairly constant value. It must be high enough to keep any water as a vapour but low enough as to not oxidise the lubricating oil. There is no real restriction on the inlet air temperature as this depends on the surrounding atmospheric temperature. It is sensible to look at the worst case, which in a hot country could easily be 38 ^{o}C. The design of a car radiator used by most manufacturers is shown in Table 4.

Table 4. Design Data for Car Radiators

Radiator Size	360 * 690 * 15 mm
Fin Pitch	1.1 mm
Materials	Aluminium Header, Fins and Tubes Plastic Tanks
Heat Transfer Area	8.664 m^2
Operating Temperature	110 ^{o}C max.
Operating Pressure	0.9 bar (typical petrol engine spec.)
Fluid Flow Rate	30 - 120 l/min
Fluid Inlet Temperature	87 ^{o}C - 95 ^{o}C
Fluid DT	4 ^{o}C - 6 ^{o}C
Fluid Pressure Drop	190 mm Hg (approx. 25 kN/m^2)

Specifications used for Case Study

The main specifications which need to be adhered to are those of the water side, i.e. the flow rate, inlet temperature, temperature drop and pressure drop. The air flow details are not as important and as long as the spiral PFCHE can perform the same duty at the same fluid flow, then it has great potential as a car radiator. The fluid used is mainly water with a small amount of anti-freeze added, this apparently does not change the physical properties of the fluid by a significant amount and so those of water have been used. The water flow chosen was 50 l/min, with an inlet temperature of 87 ^{o}C and a temperature drop of 6 ^{o}C. The air inlet temperature was 38 ^{o}C, and from these details the duty can be obtained and therefore the case study is defined.

Design Case for Spiral PFCHE

The water flow used in the actual car radiator is much greater than that used in the test rig. It was necessary, therefore, to redesign the spiral PFCHE to cope with this extra flow rate. The only way to keep down the fluid velocity, and also pressure drop, is to increase the frontal free flow area for the fluid. To cope with a flow of 50 l/min the increase has to be large. It was therefore decided to have several water channels wound together on one mandrel. This would greatly increase the fluid flow area. It was also decided to have spacers of 2 mm between the fluid sheets (previously 1 mm) and an effective corrugation amplitude of 1 mm (previously 0.5 mm), this would have little effect on the heat transfer performance while providing the extra area needed. The total length of the spiral was chosen to be 4 m by 0.1 m, these values are arbitrary but sensible. The air flow path is therefore 10 cm and the water flow path is 1 m because there are 4 water channels each taking 1/4 of the total water duty. With this information the calculations necessary were performed on Excel 4.0 using correlation's established from experimental studies. The results of the case studies are presented in Table 5.

Table 5 : The Results of the Case Study for the Car Radiator

Duty	20.4384 kW		
Total Heat Transfer Area	1.24 m^2		
Length	3.72 m		
Diameter of Spiral	19 cm		
Water Side		**Air Side**	
Total Flow rate	50 l/min	Total Flow rate	0.89 l/min
Flow rate per channel	0.203 kg/s		1.013 kg/s
Free Flow Area	1.6×10^{-3} m^2	Free Flow Area	0.00683 m^2
Velocity	0.51 m/s	Velocity	130.1 m/s
Inlet Temperature	87 ^oC	Inlet Temperature	38 ^oC
Outlet Temperature	81 ^oC	OutletTemperature	58 ^oC
DT	6 ^oC	DT	20 ^oC
Cp	4190 J/kg K	Cp	1006.6 J/kg K
k	0.66 W/m K	k	0.0265 W/m K
m	3.6×10^{-4} Ns/m^2	m	1.89×10^{-5} Ns/m^2
r	973 kg/m^3	r	1.14 kg/m^3
Prandtl Number	2.285	Prandtl Number	0.718
Reynolds Number	929	Reynolds Number	5179
$jh = 0.2145 * Re^{-0.89}$	0.001873	$jh = 0.49 * Re^{-0.52}$	0.005738
$h = G*Cp*jh*Pr^{-2/3}$	2.292 kW/m^2 K	$h = G*Cp*jh*Pr^{-2/3}$	0.589 kW/m^2 K
U(overall heat transfer)	0.469 kW/m^2 K		
Effectiveness	41%		
NTU	0.57		
Path Length	0.9 m	Path Length	0.1 m
Friction Factor	0.035	Friction Factor	0.00218
Pressure Drop	24.63 kN/m^2	Pressure Drop	12.75 kN/m^2

The case study proves that the spiral heat exchanger can, in theory, be used as a car radiator with the following dimensions:

Dimensions of a Spiral Polymer Radiator

Spiral Diameter	19 cm (including 10 cm mandrel)	
Width	10 cm	
Duty	20.38 kW	
Fluid Pressure Drop	24.63 kN/m^2	(spec. of 25 kN/m^2)
Effectiveness	41%	

The result is a very compact heat exchanger with has the potential to fit into any given space in a car. It fulfils all the specifications which the manufacturers require and has the benefit of being significantly smaller and lighter than the metallic radiators. The thermal effectiveness seems to be low but this is a result of the large temperature difference between the hot and cold fluids as compared to the low temperature drop in the hot liquid side. This is a result of the unit being studied rather than the spiral PFCHE not being suitable.

Conclusions

A compact spiral polymer film heat exchanger has the potential to be effectively used as a car radiator. It needs to be fabricated to the above mentioned specifications and tested to discover whether the assumptions hold. If the theory is correct then the rapid response of the spiral PFCHE to changes in temperature would make it ideal in the unsteady state conditions of a car radiator.

References

1. Recent Advances in Polymer Heat Exchanger, R. J. Jachuck, 1st International Conference on Science Engineering & Technology of Intensive Processing, September 1995, Nottingham, UK.

2. Process Intensification: Polymer Film Compact Heat Exchanger (PFCHE), R. J. Jachuck & C. Ramshaw, Trans IChemE, Vol. 72, Part A, March 1994.

BOILING IN A CENTRIFUGAL FIELD: THE CASE OF FILMS FORMED BETWEEN TWO CLOSELY SPACED HORIZONTAL SPINNING DISCS

S. YANNIOTIS and D. KOLOKOTSA
Agricultural University of Athens
Iera odos 75, 118 55 Athens, Greece

Abstract

An experimental unit was designed and constructed in order to study the heat transfer characteristics of liquid films formed between two closely spaced spinning discs. The heat flux vs. wall superheat was measured experimentally for rotational speeds of 200, 600, 1000 RPM and flow rates of 1, 3 and 5 l/min of distilled water and corn syrup solutions. From these results the heat transfer coefficient in each disc was calculated. The heat transfer coefficient increases as the speed of rotation and the saturation temperature increases. The speed of rotation influences the heat transfer coefficient more than the other parameters.

INTRODUCTION

Process intensification is aiming at reducing the physical size of a plant mainly by increasing the performance of the equipment. The use of centrifugal fields is one way to increase substantially the performance of heat and mass transfer equipment. The liquid films that are usually created on spinning surfaces exhibit high heat and mass transfer coefficients due to high film velocity and low film thickness.

Devices in which liquid films are created due to the centrifugal force on a rotating disc or other surfaces of revolution, are technically suitable for numerous applications. Distillation, absorption, extraction, boiling, condensation and other operations involving heat and mass transfer are susceptible to exploitation of centrifugal fields. Thermodynamic efficiency of such units can be high since they can operate with small driving forces (temperature difference, concentration difference).

A spinning disc is suitable for single phase or two phase heat transfer. The first major attempt to exploit centrifugal fields in two phase heat transfer was in the late 50's where the Hickman still was invented and used for sea-water desalination. Since then, more work on heat transfer on a spinning disc has been published (1-9).

Centrifugal evaporators that resemble the Hickman still are in industrial use today. In these evaporators the feed is pumped to the underside of a conical heating surface which rotates at 400 to 1500 rpm. Under the influence of the centrifugal force, which is around 200-400 g, the liquid is spread over the conical surface to a very thin film (about 0.1 mm) where it is heated and concentrated. The concentrate is collected at the periphery and is pumped out by means of a stationary scoop which is dipping into the rotating liquid ring. The heating medium (steam), is supplied on the top surface of the disc. The condensate is thrown away by the centrifugal force so that the heating surface is always dry.

The existing industrial centrifugal evaporators are mainly applied today in cases where quality of the product is at a premium and the necessary evaporation capacity is relatively low (less than 5000 kg water/h). Heat sensitive materials can be processed in a very gentle way, since the contact time of the product with the hot surface is less than one second, there are not hot spots on the equipment, the spread of the residence time is very narrow and the hold-up liquid content is small. Typical applications where these evaporators can be used include:

- food products such as fruit juices, herbal and seasoning extracts
- pharmaceutical products such as antibiotics, enzymes and protein solutions
- chemical products such as higher alcohols.

Because of the high initial cost of the equipment, multiple effect systems are not usually used and therefor energy efficiency in existing centrifugal evaporators is low. They are usually used as single-effect units with high temperature difference so that the product is concentrated in a single pass and retains very high quality.

Although the initial cost of these systems is higher than conventional evaporators, it could probably be justified if it was examined under the prospective of the process intensification philosophy of the whole plant. In such a case, pipes, supporting structure, foundations, installation cost and civil engineering would be lower and the total cost of the plant could become competitive. In addition, as automation in manufacturing is developed, the high cost of machining, that is required in such systems, will be reduced.

Our work on heat transfer characteristics of boiling on a spinning disc has been presented in previous communications (7-9). Smooth and grooved discs with spiral, circumferential and radial grooves were tested in an experimental unit with a single horizontal disc rotating on a vertical shaft and heated with steam from underneath. Heat transfer coefficients from about 2 kW/m^2K to 15 kW/m^2K were obtained when the rotational speed varied from 0 rpm to 1000 rpm.

There is not any report in the literature dealing with liquid film flow and heat transfer between two co-rotating discs. However, if a compact heat exchanger of industrial importance based on intensified heat transfer in a centrifugal field is to be realised, the unit should consist of a multiple matrix of discs closely spaced. In this communication we will present the heat transfer characteristics in boiling of liquid films formed between two closely spaced horizontal spinning discs.

EXPERIMENTAL WORK

The test rig

The experimental set-up that was used in this work is shown in Figure 1. The main unit (Figure 2) consists of 2 discs, 30 cm diameter and 20 mm thickness, co-rotating on a central vertical shaft. The clearance between the two discs is 5 mm. A liquid film distributor located at the centre between the two discs ensures liquid feed on both discs. A stainless steel sheet-ring, 1 mm thick and 6 cm wide, with D_i=29 cm and D_o=35 cm, is placed between the two discs (in the outer diameter) to ensure that the two films are collected separately and they are not mixed. The stainless steel ring is hold in place by two Teflon rings which are bolted on the periphery of the discs and with each other. The Teflon rings have been machined so that the liquid films are guided to two separate collection tanks which are connected to the main unit. Two stationary electric resistances supply heat (up to 4 kW each) to the non-wetted side of each disc by radiation. An electronic variac is used to adjust the voltage (power) to the resistances. A shell and tube heat exchanger is used to condense the vapour generated during the evaporation experiments. A DC electric motor with variable speed drive is used to rotate the disc. A liquid ring vacuum pump is used to remove the non-condensables from the system. Mechanical seals are used where the liquid flows from non-rotating to rotating parts in order to operate the system under vacuum. Concentrate holding tanks and a vapour condensate holding tank are connected to the main unit. The metallic parts of the system are made of stainless steel. The main unit has two glass windows to allow visual observation. The unit is strippable.

Instrumentation

The data acquisition system consists of two peripheral stations connected to a PC for central control and storage of the data. Station 1, which is attached at the bottom end of the rotating shaft, is used for the temperature measurement at several points on the discs. It is equipped with a wireless optical transmitter (instead of slip rings) which transfers the data to a stationary receiver located about 10 cm underneath (photo 1) and from there to the PC.

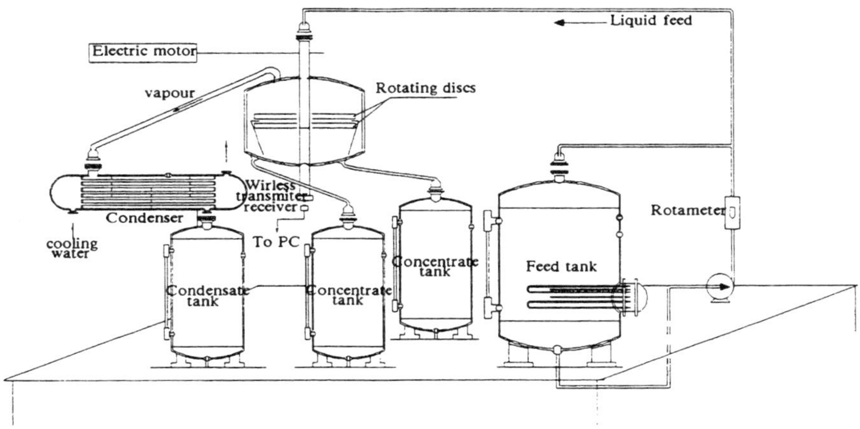

Figure 1. Experimental set-up

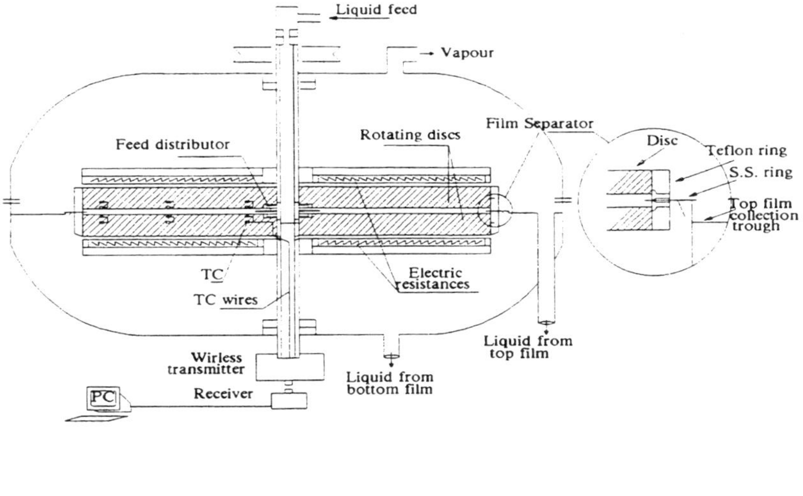

Figure 2. Main unit

Station 2, which is stationary, is used for the measurement of the temperature and the flow rate of the fluids. Every 60 sec all the measurements are read. A separate controller is used in order to adjust manually the rotational speed of the discs to within 10 rpm.

The discs surface temperatures are measured with six Chromel-Alumel thermocouples of 0.5 mm wire diameter with an accuracy of 0.1°C. Three thermocouples were placed on each side of the discs so that the average temperature of three zones of the disc of approximately equal area is measured. The inlet and outlet temperatures of the fluids are measured with Chromel-Alumel thermocouples with an accuracy of 0.1°C. The flow rate of the working liquid and the flow rate of the cooling water in the condenser, are measured with two rotameters.

Experimental operation

The liquid feed is introduced in the centre of the discs, through the hollow shaft. After flowing across the co-rotating discs as two separate thin films, the liquid is collected in the two concentrate tanks (one for each disc). The generated vapour is taken to the condenser. The vapour condensate and liquid concentrates are collected and measured in the tanks. The system operates in this way until steady state is reached. Then the experiment lasts for about twenty minutes. During this time all the measurements are recorded by the PC for later processing.

Analysis of data

The heat flux for each zone on the discs is calculated knowing: a) the temperature difference between the top and the bottom surface of the disc, as measured with the 2 thermocouples located in each zone of the disc, b) the distance between the thermocouples in the direction of heat flow and c) the thermal conductivity of the aluminium (140 W/mK, as given by the supplier of the aluminium). The heat transfer coefficient for each zone is calculated from the heat flux of each zone and the wall superheat (the temperature difference between the top surface of the zone and the vapour temperature). The average heat transfer coefficient for the whole surface of the disc is calculated as the weighted average of the heat transfer coefficients of the three zones on the discs (the weight for each coefficient was the fraction of the zone area to the total area of the disc).

Photo 1. Data logger with the wireless transmitter/receiver

RESULTS AND DISCUSSION

The results of this work showed that the proper distribution of the liquid over the surface of a horizontal smooth flat single disc or between two similar discs does not impose a difficult problem on spinning disc technology. A liquid film is formed on the bottom surface of the top disc as well as on the top surface of the bottom disc at least for the flow rates used in our experiments. This argument is supported by: a) the liquid from each disc was collected in a different tank and thus it was possible to quantify the liquid distribution between the top and the bottom disc and b) the observation that local overheating on the surface of the disc (implying dry spots) was not observed at any point and time.

The heat transfer coefficient for both, the top and the bottom disc was the same. It ranged from about 5 kW/m^2K to about 13 kW/m^2K when the speed of rotation was between 200-1000 rpm, the wall superheat between 1-4 K and the boiling temperature 40°C. In general, it can be stated that for the same temperature difference the heat flux increases when the speed of rotation increases. The heat transfer coefficient increases as the rotational speed increases.

The heat transfer coefficient was measured for 1, 3 and 5 l/min feed flow rate (Figure 1). At 3 and 5 l/min the heat transfer coefficient was the same or the difference was smaller than the experimental error and could not be detected. However as the flow rate was decreased to 1 l/min the heat transfer coefficient increased slightly. It should be noted here that the above flow rates refer to the total flow rate which was distributed between the two discs.

The above experiments were run under vacuum (the boiling temperature was equal to 40°C). When the saturation temperature increased from 40°C to 60°C, the heat transfer coefficient increased by about 25%.

The experimental results reported above were obtained using pure de-ionised water as liquid to be evaporated (evaporant). In a number of experiments 45° and 60°Brix corn syrup was used in order to see the effect of the physical properties (mainly viscosity) of the evaporant on the heat transfer. The heat transfer coefficient in this case varied between 2-5 kW/m^{2}K for rotational speeds between 600-1000 rpm, flow rate 3 l/min and boiling temperature 40°C.

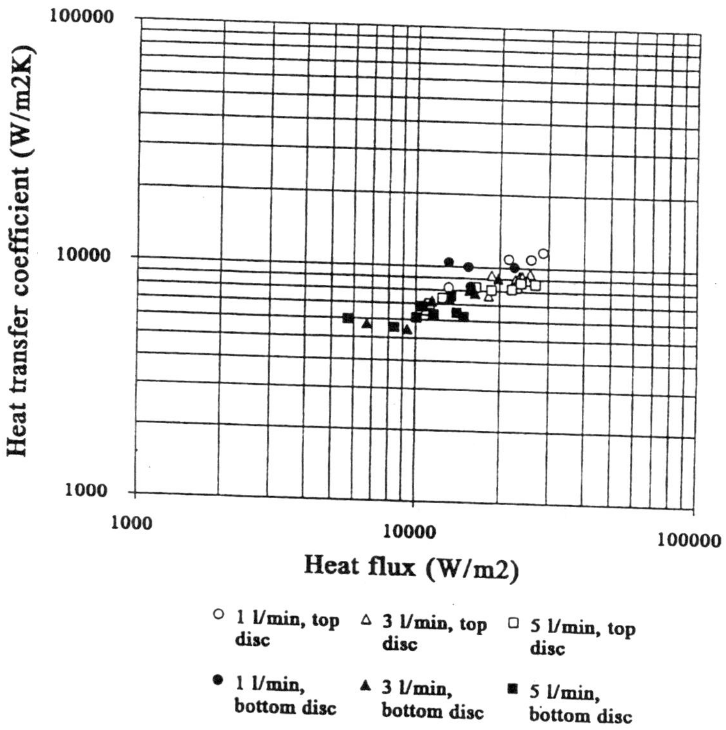

Figure 1. Heat transfer coefficient for the top and bottom disc at 600 rpm and 40°C saturation temperature.

REFERENCES

[1] BELL C. The hydrodynamic and heat transfer characteristics of liquid films on a rotating disc. Ph.D. Thesis, University of Newcastle upon Tyne, 1975.

[2] HICKMAN K. C. D. Centrifugal Boiler Compression Still. Ind. Eng. Chem., vol. 49, 786-789, 1957.

[3] JACHUCK R. J. J. and RAMSHAW C. Heat transfer characteristics of tailored rotating surfaces. EUROTHERM SEMINAR NO 26, Hariot-Watt Univ. Edinburgh, Scotland, Sept. 1993.

[4] KAHN J. R. Heat Transfer on a Rotating Surface with and without Phase Change, Ph.D. Thesis, Univ. of Newcastle Upon Tyne, 1986.

[5] RAMSHAW C. The opportunities for exploiting centrifugal fields. Heat recovery systems & CHP, vol. 13, 493-513, 1993.

[6] WOOD R.M. and WATTS B. E. The flow, heat and mass transfer characteristics of liquid films on rotating discs. Trans. ICHE vol. 51, 315-322, 1973.

[7] YANNIOTIS S., KOLOKOTSA D. and VALACHIS P. A rotating disc heat exchanger. In Energy Efficiency in Process Technology, edited by P. Pilavachi, Elsevier Applied Science, 613-623, 1993.

[8] YANNIOTIS S., KOLOKOTSA D. AND VALACHIS P. Heat transfer enhancement on rotating surfaces. EUROTHERM SEMINAR No 33, Paris, France, Oct. 1993.

[9] YANNIOTIS S. AND KOLOKOTSA D. Experimental study of water vapour condensation on a rotating disc. Accepted for publication in the International Communications in Heat and Mass Transfer, 1995.

ACKNOWLEDGEMENTS

The financial support offered by CEC (JOULE I program) and the Centre of Renewable Energy Sources is gratefully acknowledged. The help of Mr M. Mpeladakis in carrying out the experiments is also acknowledged.

DESIGN METHOD FOR PLATE EVAPORATORS AND CONDENSERS

B. THONON
GRETh , CEN Grenoble, FRANCE

INTRODUCTION

The first applications for plate heat exchangers were for milk pasteurisation and afterward for heat recovery in several industrial processes, but essentially in single-phase flows [1] and [2]. Classical plate heat exchangers are gasketed, and their range of application is limited by pressure or temperature, but with brazed or welded plate heat exchangers temperature from -180°C to + 500°C and pressure up to 30 bars can be reached. This allows to use plate heat exchangers in industrial processes either as evaporators or as condensers [3], [4] and [5], brazed or welded plate heat exchangers are also used in ammonia cooling units [6] and [7] and in automotive air conditioning [8] and [9]. Some large units are used in catalytic reforming processes [10] and [11].

These heat exchangers have high heat transfer performances, and allow a significant size reduction. Therefore, as they are compact, the liquid hold-up is smaller than in conventional heat exchangers. As plate heat exchangers works in parallel flow (co or countercurrent), the temperature enthalpy diagram method can be applied to the selection of the optimal number or size of the channel, and to assess the design method there is a need of a good knowledge of the heat transfer and pressure drop laws in corrugated channels.

PLATE HEAT EXCHANGER DESCRIPTION

Heat exchanger geometry

A plate heat exchanger consists of a pack of corrugated plates, which creates the channels (Figure 1). The hot fluid and the cold fluid circulate in co or countercurrent in the different channels (Figure 2). Multipass plate heat exchangers exist, but for two-phase flow processes it is recommended to use single-pass arrangement on the phase change side.

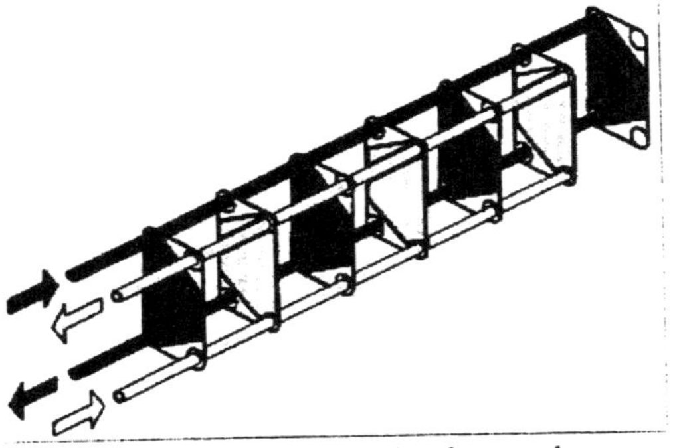

Figure 1 : Schematic view of a plate heat exchanger

Figure 2 :Countercurrent flow

Channel geometry
The channel is created by assembling two corrugated plates, and the way of packing the plates leads to the fact that the mean gap between two plates is equal to the height of the corrugation. The plates are characterised by the length and the width of the heat transfer surface, but also by the corrugation angle (figure 3).

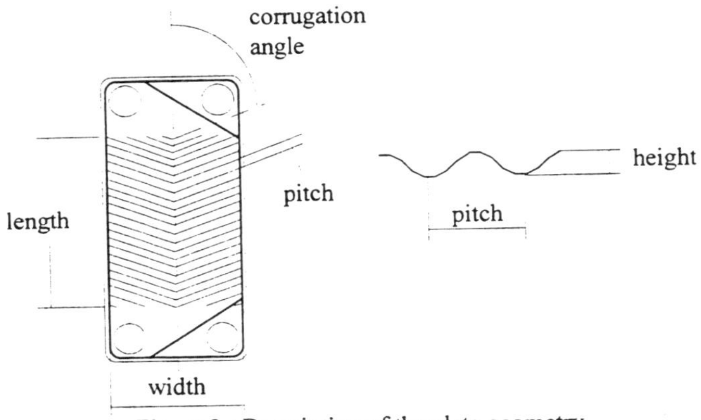

Figure 3 : Description of the plate geometry

DESIGN METHOD
Introduction
The goal of a design method is to find the geometry which gives the required heat duty for an acceptable pressure drop, and this with the lowest heat transfer surface. If we follow the evolution of the heat rate and pressure drop in function of the number of channel, three cases exist (Figure 4) :

- Zone A : The heat exchanger is oversized and there is no pressure drop limitation.
- Zone B : The heat exchanger is undersized and there is no pressure drop limitation.
- Zone C : The heat exchanger is undersized and there is a pressure drop limitation.

The optimal geometry should be located at the border between zone A and zone B.
As the heat exchanger geometry depends not only on the number of channels, but also on the corrugation angle and on the plate size, a software is required if we want to test all the different possible geometries.

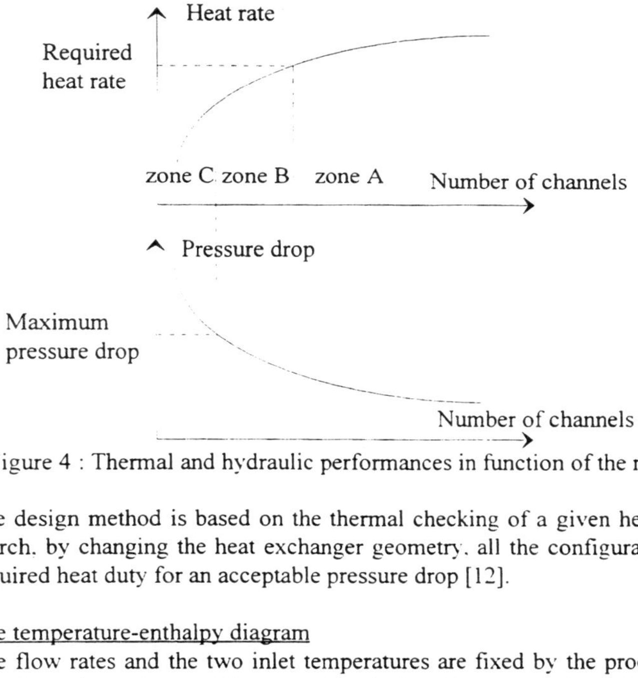

Figure 4 : Thermal and hydraulic performances in function of the number of channels

The design method is based on the thermal checking of a given heat exchanger, and to search, by changing the heat exchanger geometry, all the configurations which give the required heat duty for an acceptable pressure drop [12].

<u>The temperature-enthalpy diagram</u>
The flow rates and the two inlet temperatures are fixed by the process, and for a given heat duty, the outlet conditions can be calculated by an enthalpy balance.

$$\dot{M}_1 \, (h_{1\,in} - h_{1\,out}) = \dot{M}_2 \, (h_{2\,out} - h_{2\,in}) \qquad (1)$$

The knowledge of the physical properties of each fluids leads to the buildings functions relating the enthalpy to the temperature.

$$h = \mathbf{f}(T) \qquad T = \mathbf{f}^{-1}(h) \qquad (2)$$

As we know the inlet and outlet thermal condition, the energy balance between the inlet and any point x inside the exchanger is given by :

for a cocurrent heat exchanger

$$\dot{M}_1\,(h_{1\,in} - h_{1\,x}) = \dot{M}_2\,(h_{2\,x} - h_{2\,in}) \tag{3}$$

and for a countercurrent heat exchanger.

$$\dot{M}_1\,(h_{1\,x} - h_{1\,out}) = \dot{M}_2\,(h_{2\,x} - h_{2\,in}) \tag{4}$$

Since these relations are valid for all points inside the heat exchanger, the enthalpy evolution of one fluid can be related to the other one. Therefore the temperature enthalpy diagram may be determined through the **f** functions (Figure 5).

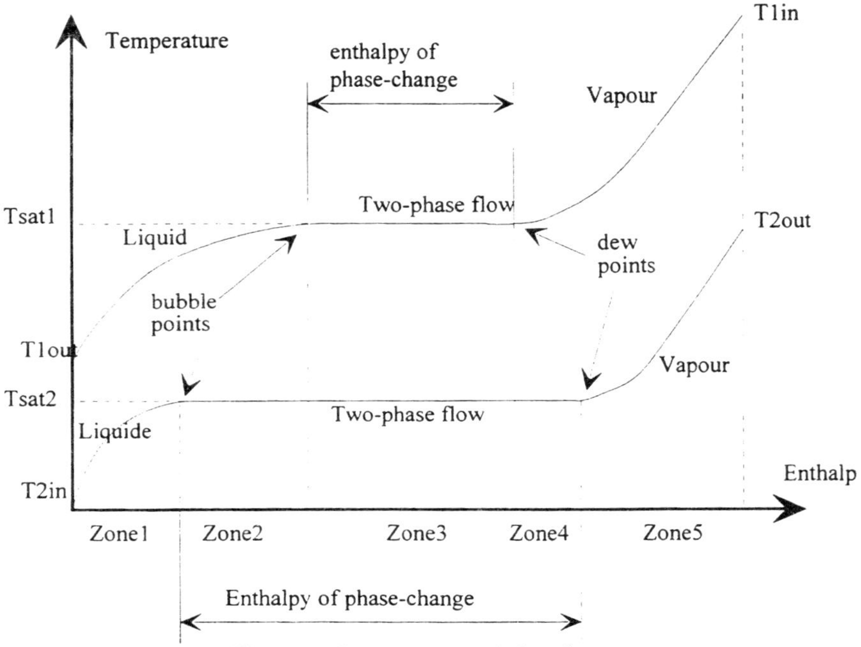

Figure 5 : Temperature/enthalpy diagram

<u>Segmentation of temperature/enthalpy diagram</u>

In single phase flows, the heat transfer coefficients are relatively constant along the heat exchanger, and the segmentation can be chosen without any flow constraint. But in two-phase flows, there are some strong variations of the heat transfer coefficient near the dew and bubble points. In consequence, these particularly points needs to be consider as borders for the segmentation (figure 5).

The checking mode

The evolution curve is segmented in an appropriate number of cells. Then in each cell, limited by the point x and x', the local energy balance is expressed :

$$\dot{M}\,(h_x - h_{x'}) = U\, A_{xx'}\, \Delta T\, ln \qquad (5)$$

Where U is the overall heat transfer coefficient and $A_{xx'}$ the heat transfer surface.
The mean log temperature difference is given by :

$$\Delta T\, ln = \frac{(T_{1x} - T_{2x}) - (T_{1x'} - T_{2x}')}{Ln \dfrac{T_{1x} - T_{2x}}{T_{1x'} - T_{2x'}}} \qquad (6)$$

The overall heat transfer coefficient can be evaluated from the local heat transfer coefficients of both fluids.

$$U = \left(\frac{1}{\alpha_1} + \frac{1}{\alpha_2} + \frac{e_w}{\lambda_w} + R_f \right)^{-1} \qquad (7)$$

Since these quantities (α_1 and α_2) are known or can be evaluated from correlations, the required heat transfer surface $A_{xx'}$ can be calculated. The required heat transfer surface A_r is determined by the integration of (5) from the inlet to the outlet of the heat exchanger. The actual heat transfer surface A_c is given by the number of channels and by the length and the width of the plates. Then, the ratio of the required and actual heat transfer surface characterises the thermal sizing of the heat exchanger.

Great care is to be taken in the choice of the fouling resistance R_f. It has been shown that plate heat exchangers are less prone to fouling than conventional shell and tube heat exchanger due to the higher level of the shear stress [13]. Furthermore, if the TEMA values are applied, the heat exchanger will be overdesigned, and it may lead to poor operating conditions. It is often recommended that the fouling margin should not exceed 25% of extra surface.

But to assess the design method there is a need of heat transfer and pressure laws, for the different flow conditions (single-phase, evaporation and condensation).

HEAT TRANSFER

Single phase flow

Extensive information and literature have been summarised by Raju and Bansal [1] and Shah and Focke [2]. These previous works have shown that the major geometric parameter is the corrugation angle. For given flow conditions, the heat transfer coefficient can be 6 times greater than in a comparable plain channel.
Some correlation have been established in function of the corrugation angle, but they cannot be generalised to give a high degree of prediction because the shape of the corrugation and the operating conditions have some influence [14].

The general form of these correlation is :

$$\mathrm{Nu} = a\ \mathrm{Re}^b\ \mathrm{Pr}^c = \frac{\alpha\ d_h}{\lambda} \tag{8}$$

With $0.1 < a < 2.$
$0.6 < b < 0.8$
$0.3 < c < 0.4$

The Reynolds and Prandtl numbers are defined by :

$$\mathrm{Re} = \frac{\dot{m}\ d_h}{\mu} \qquad \mathrm{Pr} = \frac{\mu\ c_p}{\lambda}$$

Generally, the plate heat exchanger manufacturers have a good knowledge of the thermal performances in single phase flow.

Evaporation

In such corrugated channels, annular flow pattern dominates in a large range of applications, unless the vapour quality is low [8] and [15]. This flow pattern, in evaporation, allows a high heat transfer coefficient. As in tube, two phenomena occurs : nucleate boiling (nb) and convective boiling (cv). Ohara [8] and Marvillet [16] have observed that nucleate boiling have no influence, unless for high heat flux or low mass flux.

In nucleate boiling, the heat transfer coefficient is function of the heat flux and of fluid properties. The Cooper correlation [17] is used to estimated the nucleate boiling term :

$$\alpha_{nb} = 55\ \tilde{M}^{-0.5}\ p^{*0.12}\ \dot{q}^{2/3}\ (-\log_{10} p^*)^{-0.55} = \frac{\dot{q}}{T_w - T_{sat}} \tag{9}$$

To estimate the convective term, we use an enhancement factor F, which characterises the heat transfer enhancement due to the liquid vapour interaction.

$$\alpha_{cv} = F\ \alpha_L \tag{10}$$

Where the liquid heat transfer coefficient is calculated with an appropriate correlation for corrugated channels.

$$\alpha_L = a \left(\frac{\dot{m}\,(1-x)\ d_h}{\mu_L} \right)^b \left(\frac{\mu_L\ c_{pL}}{\lambda_L} \right)^c \frac{\lambda_L}{d_h} \tag{11}$$

The enhancement factor F is calculated with the Chen correlation, which was originally developed for plain tubes.

$$F = 1 + \frac{1.8}{Xtt^{0.79}} \tag{12}$$

Where Xtt is Lockhart-Martinelli number :

$$Xtt = \left(\frac{1-x}{x} \right)^{0.9} \left(\frac{\rho_G}{\rho_L} \right)^{0.5} \left(\frac{\mu_L}{\mu_G} \right)^{0.1} \tag{13}$$

The two phase heat transfer coefficient is taken as the greater value of the nucleate boiling term and the convective boiling term.

$$\alpha = \mathrm{Max}(\alpha_{nb}, \alpha_{cv})$$

The critical dryness quality, above the one the heat transfer coefficient drops, is at lower values than in plain tubes, typically $x_c \approx 0.7$. For a pure vapour, the heat transfer coefficient can be estimated from single phase correlation, thus the two phase heat transfer coefficient is estimated by interpolation between the value at $\alpha_{x_c=0.7}$ and the value of a pure vapour α_{G0}.

$$\alpha_{(x>0.7)} = \alpha_{(x=0.7)} - \frac{x - x_c}{1 - x_c}\left(\alpha_{(x=0.7)} - \alpha_{GO}\right) \tag{14}$$

With

$$\alpha_{G0} = a\left(\frac{\dot{m}\, d_h}{\mu_G}\right)^b \left(\frac{\mu_G\, c_{pG}}{\lambda_G}\right)^c \frac{\lambda_G}{d_h} \tag{15}$$

This approach was originally used by Cornwell [18], and it seams to give estimation of our data. Table 1 summarises the results for several type of plate heat exchangers.

Geometry	Fluid	Heat flux kW/m^2	Mass velocity kg/m^2 s	Validity % in ± 30%
Plain [16]	R 22	10-60	200-600	77 %
Corrugated [14]	R 22	5-60	200-900	85 %
Enhanced	R 113	3-16	70-600	70 %

Table 1 : Comparison with experimental results

Condensation

Some local measurements of the heat transfer coefficient [4] and [19] have clearly shown the existence of two regimes of condensation depending on the mass flux :

- At low mass flux ($\mathrm{Re}_L < 300 - 1000$) : a gravity controlled regime, where the liquid flow may be separate from vapour flow.
- At high mass flux : a shear controlled regime, where a turbulent flow model must be applied

For the gravity controlled regime, the Nusselt theory underpredicts by a factor 2 to 3 the heat transfer coefficient [4] and [20]. This difference may comes from the fact that the liquid film is drained in the furrow of the corrugations.

In the shear controlled regime, several correlations exist based on heat-momentum analogy, and one of the most common is the Boyko-Kruzilihn correlation. The heat transfer coefficient in condensation is reported to heat transfer coefficient of the liquid.

$$\alpha = \alpha_{L0}\left[1 + x\left(\frac{\rho_L}{\rho_G} - 1\right)\right]^{0.5} \qquad (16)$$

With

$$\alpha_{L0} = a\left(\frac{\dot{m}\, d_h}{\mu_L}\right)^b \left(\frac{\mu_L\, c_{pL}}{\lambda_L}\right)^c \frac{\lambda_L}{d_h} \qquad (17)$$

This correlation has been tested on several heat exchangers and seems to give some good results for condensation of water steam [21], but for refrigeration application it was found to be less reliable. The heat transfer coefficient is underestimated for low corrugation angle, but for high corrugation angle the heat transfer coefficient is overestimated.

PRESSURE DROP

Single phase flow

The pressure drop, in corrugated channels, can be increased by a factor one hundred compared to a equivalent plain tube, and this comes from the higher level of turbulence in the channel. The pressure drop is calculated from :

$$\Delta p = 4\, f\, \frac{\dot{m}^2}{2\,\rho}\, \frac{L}{d_h} \qquad (18)$$

Where f is the channel friction factor, which is a function of the Reynolds number and which depends on the corrugation angle. For a Reynolds number of 1000., the increased of the friction factor is : $f_{30°} / f_{plain} = 9.2$ and $f_{60°} / f_{plain} = 83$

It must be noted that for a Reynolds number of 1000, the flow is in laminar regime in a plain channel, while it is in turbulent regime for corrugated channels.

Two phase flow

The knowledge of the pressure drop, during evaporation or condensation, cannot be dissociated from the heat transfer. On one hand, the pressure drop will give the mechanical energy needed to flow the process fluid. On the other hand, the saturation temperature depends on the actual pressure. As the pressure drop in corrugated channels can be important, the saturation temperature can varies of several degrees between the inlet and the outlet, and affect the mean temperature difference between the cold and hot fluid.

In two-phase flows, as in single phase flow, there a significant increase of the pressure drop compared to a plain tube. The two-phase flow pressure drop is estimated from the pressure drops of the liquid and gas flows.

$$\Delta p = \Delta p_L + C\sqrt{\Delta p_L\, \Delta p_G} + \Delta p_G \qquad (19)$$

Where the liquid and gas pressure drop are calculated from the single-phase flow correlations:

$$\Delta p_L = 4\, f_L \frac{\dot{m}^2\,(1-x)^2}{2\,\rho_L} \frac{L}{d_h} \tag{20}$$

$$\Delta p_G = 4\, f_G \frac{\dot{m}^2\, x^2}{2\,\rho_G} \frac{L}{d_h} \tag{21}$$

For plain tubes, the C constant depends on the flow conditions and varies from 5 to 20. For corrugated channel the value C=8 allows a good estimation of the two-phase flow pressure drop [14] (for a tube bundle C=8).

CONCLUSION

In single-phase flows, the overall performances of plate heat exchangers can be estimated with a good accuracy by the proposed model. But in some critical working conditions, complex pass arrangements, some more sophisticated models needs to be used to take into account flow maldistribution or edge effects [22].

In two-phase flow applications (evaporation or condensation), the proposed model can be used for the design of plate heat exchangers, but the accuracy will depend on reliability of the heat transfer and pressure drop correlations. Some more work is needed to generalise the heat transfer coefficient correlations in function of the geometrical and operating parameters, this is particularly true in condensation [23]. Furthermore, a perfect flow distribution is assumed in the model, but in evaporation processes the liquid/vapour mixture is not uniformly distributed between the channels, and this will affect the overall performances.

Industrial process fluids are often mixtures and the temperature-enthalpy design method is still valid as far as the boiling or condensation curve can be established. But some experimental work is needed to develop and validate models to estimate the heat transfer coefficients for boiling and condensation of mixtures in complex geometry.

NOMENCLATURE

A	heat transfer surface	m^2
C	constant (eq. 17)	
c_p	heat transfer capacity	J/kg K
d_h	hydraulic diameter	m
e	plate thickness	m
F	enhancement factor	
f	friction factor	
h	enthalpy	J/kg
L	length	m
$\dot{M}$	mass flow rate	kg/s
$\tilde{M}$	molar mass	g/mol
$\dot{m}$	mass flux	kg/s m^2
p	pressure	Pa
p^*	reduce pressure	Pa/Pa
$\dot{q}$	heat flux	W/m^2
R_f	fouling resistance	m^2 K/W
T	temperature	K
U	overall heat transfer coef.	W/m^2 K
x	quality	
α	local heat transfer coef.	W/m^2 K
λ	thermal conductivity	W/m K
μ	dynamic viscosity	Pa s
ρ	density	kg/m^3

<u>subscripts</u>

c	actual surface
cr	critical
cv	convective boiling
G	gas
G0	gas only
in	inlet
L	liquid
L0	liquid only
nb	nucleate boiling
out	outlet
r	required surface
sat	saturation
w	wall
1	hot fluid
2	cold fluid

<u>Non dimensional numbers</u>

Nu	Nusselt number
Pr	Prandtl number
Re	Reynolds number
Xtt	Lockhart-Martinelli number

REFERENCES

[1] Raju K.S.N. et Bansal J.C. "Plate heat exchanger design", in Low Reynolds Number Flow Heat Exchanger, Hemisphere Publishing Corporation, 1983

[2] Shah R.K. et Focke W.W. "Plate heat exchanger and their design theory", in Heat Transfer Equipment Design, Hemisphere Publishing Corporation, 1988

[3] Syed A. "The use of plate heat exchangers as evaporators and condensers in process refrigeration", Heat Exchange Engineering, E.A. Foumeny et P.J. Heggs ,Ellis Horwood Series in Chemical Engineering, Volume 1 Design of Heat Exchangers, Chapter 10, 1992

[4] Kumar H. "The design of plate heat exchangers for refrigerants", Proceedings of the Institute of Refrigeration 1991-92, 5-1 to 5.5, 1992

[5] Panchal C.B., Hillis D.L. et Thomas A. "Convective boiling of ammonia and freon R22 in plate heat exchanger", ASME/JSME Thermal Engineering Conference Honolulu, pp 261-268, 1983

[6] Wand S. "Practical design tips for plate heat exchangers in ammonia refrigerants systems", Proceedings of the IIAR 16th Annual Meeting, St Louis, Missouri, 1994

[7] Young M. "Plate heat exchangers as liquid cooling evaporators in ammonia refrigeration systems", Proceedings of the IIAR 16th Annual Meeting, St Louis, 1994
[8] Ohara T., Yamamoto T. and Fujita H. "Heat transfer and pressure drop of boiling flow in cross-ribbed flat channels", Int Com of Heat and Mass Transfer, Vol 17, n°5, pp 556-557, 1990
[9] Marvillet C. "Recent developments in heat exchanger for automotive applications", in Recent developments in finned tube heat exchanger,DTI, Energy Technology, Denmark, 1993
[10] Fauconier J.C., Gorenflo F. and Thomas R., "A shell and plate heat exchanger for the catalytic reforming unit at OMW refinery Karlsruhe", in Heat Exchanger Technology : Recent Developments, EETI, France, 1994
[11] Grillot J.M. "Liquid side fouling of compact heat exchangers : first and second periodic reports.", EEC Joule contract, JOUE-CT92-0033, 1994
[12] Vidil R., Ratel G. et Grillot J.M. "Thermal hydraulic performance of plate and frame heat exchangers - The CEPAJ Software", in Design and Operation of Heat Exchangers, Springer Verlag, 1992
[13] Thonon B., Grillot J.M. and Vidil R., "Liquid side fouling of plate heat exchangers"; proceedings of Fouling mitigation of industrial heat exchangers, June 1995, Shell Beach, California, 1995
[14] Thonon B., Vidil R., and Marvillet C. "Recent research and developments in plate heat exchangers", Journal of Enhanced Heat Transfer, Vol 2, n° 1-2, pp 149-155, 1995
[15] Cohen M. and Carey V.P. " A comparison of the flow boiling performance characteristics of partially heated cross-ribbed channels with different rib geometries", Int J of Heat & Mass Transfer, Vol 32, n°12, pp 2459-2474, 1989
[16] Marvillet C. "Welded plate heat exchangers as refrigerants dry-ex evaporators", in Design and Operation of Heat Exchangers , Springer-Verlag, 1992
[17] Cooper M.G. "Saturation nucleate pool boiling - A simple correlation", IChemE Symposium Series, n° 86, pp 785-793, 1984
[18] Cornwell K. and Scoones D.S. "Analysis of low quality boiling on plain and low finned tube", Proceedings of the 2nd UK Heat Transfer Conference, IMechE-IChemE, volume 1, 1988
[19] Navarro J.M. and Bailly A. "Compact brazed plate heat exchangers", in Heat Exchanger Technology : Recent Developments, EETI, France, 1994
[20] Nakaoka T et Uehara H. "Performance test of a shell and plate type condenser for OTEC", Experimental Thermal and Fluid Science, Vol 1, pp 275-281, 1988
[21] Tovazhnyanski L.L. et Kapustenko P.A. "Intensification of heat and mass transfer in channels of plate condensers", Chem Eng Communication, Vol 31, pp 351-366, 1984
[22] Thonon B., Mercier P. and Feidt M. "Flow distribution in plate heat exchangers and consequences on thermal and hydraulic performances", in Design and Operation of Heat Exchangers, Springer Verlag, 1992
[23] Thonon B. "Echangeurs à plaques : dix ans de recherche au GRETh", Revue Générale de Thermique, n°397, pp77-90, 1995

MASS

TRANSFER

COMPARISONS BETWEEN MOTIONLESS MIXERS AND EJECTORS ON THEIR GAS-LIQUID MASS TRANSFER PERFORMANCE

Z M Zhu

BHR Group Limited, Cranfield, Bedfordshire, MK43 0AJ, UK

ABSTRACT

Motionless mixers and ejectors are high intensity in-line mixing devices. They are suitable as chemical reactors for intensifying fast gas-liquid reactions and can bring the usual benefits of process intensification. This paper compares mass transfer performance of these two devices on the basis of experimental data obtained at a testing rig close to industrial scales. It is shown that mass transfer coefficients in these devices are high, about 100 times those in stirred tanks. Both similarities and differences have been found between the two devices with respect to the effects of flow parameters and spacial distribution of mass transfer coefficient. This paper also gives an example of intensification from a stirred tank to a high intensity in-line mixer for a simple fast reaction. It is shown that the use of an in-line mixer reduces the reactor volume by up to 100 times and can potentially increase the process rate. However, the in-line mixer requires much higher power, several times that of the stirred tank.

1. INTRODUCTION

Many industrial gas-liquid reactions are intrinsically fast. They can be found in the categories of oxidation, chlorination, hydrogenation, etc. In industrial plants these fast reactions are often carried out in stirred tank reactors. Because the mixing intensity of a stirred tank is relatively moderate (energy dissipation rates are typically below 5 W/kg), the reaction rate can be controlled by interfacial mass transfer. In addition, the stirred tank is a back-mixed reactor (for liquid phase). This flow pattern is undesirable for complex reactions where the intermediate is the required product. For these applications, the use of a high intensity in-line mixer as reactor can not only increase reaction rate but also increase product selectivity.

Motionless mixers and ejectors are the two main commercially available, high intensity in-line mixers suitable for intensifying fast gas-liquid reactions. The traditional uses of these devices are not as gas-liquid reactors. Motionless mixers were originally developed for liquid blending and ejectors were first used for gas compression (due to its pumping ability, see Section 2). Perhaps due to this reason, design data for these mixers as chemical reactors are quite limited in the open literature. A literature review (ref. 1) shows that for motionless mixers, only a few publications were made by researchers (ref. 2,3,4). For ejectors, most of the publications are for its compression performance (e.g. ref. 5), with only limited studies on its mass transfer (ref. 6,7).

This paper presents an experimental study of both motionless mixer and ejector on their mass transfer performance.

2. HOW DO THEY WORK?

Motionless mixers are static inserts in a pipe. An example of these is Kenics mixer shown in Figure 1. The mixer consists of a number of elements, which divide and twist the flow passing through the pipe. In turbulent flow these actions generate a high turbulence level and in the presence of a gas phase, break the gas into small bubbles and form a fine gas-liquid dispersion. Because of the high turbulence level and creation of small bubbles, specific interfacial area and mass transfer coefficient are high.

Figure 2 shows a schematic diagram of a gas-liquid ejector. The ejector consists of a nozzle, gas chamber, and mixing tube. The nozzle provides a high velocity liquid jet which creates suction at the gas chamber and entrains the gas phase into the device. The two phases experience an intensive mixing zone in the early part of the mixing tube. This intensive mixing zone is a result of rapid dissipation of the kinetic energy of the liquid jet. A fine dispersion of bubbles forms after the mixing zone.

Both devices have near plug flow characteristics for both gas and liquid phases, therefore they are desirable for complex reactions where the intermediate product is the desired one. There is also a major difference between the two devices. In a motionless mixer, both liquid and gas can be the motive fluids. Therefore both pump and compressor are usually needed. In an ejector however, only the liquid is the motive fluid. Therefore a compressor is not necessarily needed.

3. EXPERIMENTAL

3.1. Experimental set-up

An experimental rig was set up to study high intensity in-line mixers operating in vertical downflow mode. Figure 3 shows the key elements of the rig and their layout. Liquid was pumped from a 2.5 m^3 supply tank to a vertical test section where a mixer was installed. Gas was supplied from either an air compressor or nitrogen bottles and side-streamed into the mixer. The two-phase dispersion from the mixer then went to a receiving tank where the two phases were separated.

The rig had the capacity of over 10 m^3/hr, which is close to some industrial operations.

3.2 Mixer details

For study of motionless mixers, a Kenics mixer was chosen. The following summarises the geometric details of the Kenics mixer and ejector used in this study.

<u>Kenics mixer (Figure 1)</u>

A 0.04 m diameter, 6-element Kenics mixer (V=0.462 litre) was used. The mixer is the standard KMA type supplied by Chemineer Limited. Each element is 1.5 times the diameter. The mixer was housed in a transparent plastic tube to allow for visual observation of the two-phase flow.

<u>Ejector (Figure 2)</u>

The ejector was built in-house. It had a straight mixing tube (without a diffuser) and a helical spinner placed before the nozzle. The spinner had a shape similar to one element of the Kenics mixer. It gives a tangential velocity to the liquid jet and prevents a long, coherent jet to be formed in the mixing tube. The dimensions of the ejector are:

Nozzle diameter D_n = 0.0119 m
Mixing tube diameter D_m = 0.029 m
Mixing tube length L_m = 0.70 m
Mixing tube volume V = 0.452 litre
Spinner swirl number Sw = 0.018

where Sw is defined as the ratio of tangential velocity of the jet to its axial velocity at the tip of the nozzle.

3.3 Measurement of k_La

According to the Two-Film Theory, a solute transferring from one phase to another needs to overcome resistance in both liquid and gas films. For many industrial applications, the main resistance to mass transfer is in the liquid film, in which case the solute transfer rate is controlled by the liquid-side volumetric mass transfer factor (k_La)

and the driving force (C^*-C).

In this study, measurement of k_La was made with an oxygen desorption method by using nitrogen and water. Water was saturated with air in the supply tank. When nitrogen was introduced into the mixer, oxygen in the water was gradually stripped into the gas phase. Dissolved oxygen concentration in water was measured with miniature DO probes. The probes had tips 3 mm in diameter and were inserted into the gas-liquid stream for measurement.

k_La was calculated with the following equation assuming plug flow for both liquid and gas phases:

$$k_La = \frac{Q_L\,(C_1 - C_2)}{V\,\dfrac{(C_1^* - C_1) - (C_2^* - C_2)}{\ln\left(\dfrac{C_1^* - C_1}{C_2^* - C_2}\right)}} \qquad (1)$$

The equilibrium concentrations, C_1^* and C_2^*, were calculated by taking into account changes in both pressure and oxygen concentration in gas (due to oxygen stripping) along the mixers.

Temperature of water was controlled at 20 °C, using a combination of immersion heaters and a cooling coil in the supply tank.

4. VOLUMETRIC MASS TRANSFER COEFFICIENT k_La

4.1 Effect of flow parameters

Figures 4 and 5 show k_La data obtained with the Kenics mixer and ejector respectively. Each graph plots k_La against volumetric gas/liquid flow ratio at two different liquid flow rates. For both devices, an increase in liquid flow rate at constant flow ratio increased k_La. The reason for this is that increasing liquid flow rate increased the turbulence level in the liquid phase and hence reduced the resistance to mass transfer and increased k_L. The increased turbulence level also produced smaller bubbles and hence the specific interfacial area, a, was also increased.

The effect of gas flow rate on k_La was not quite the same for the two devices. For the Kenics mixer, increasing gas flow rate consistently increased k_La. For the ejector

however, k_La initially increased with an increase of gas flow rate but levelled off when gas flow rate reached a certain value. The point where k_La levelled off depended on liquid flow rate. The higher the liquid flow rate, the later (i.e. at a higher flow ratio) k_La levelled off.

This ejector's feature that k_La became insensitive to an increase in gas/liquid flow ratio when the ratio exceeded a certain level can be explained as follows. At low gas flow rates, the ejector was operating normally, i.e. the liquid entrained the gas, produced a intensive mixing zone and a gas dispersion. Increasing gas flow rate increased the fraction of gas in the device (gas holdup) hence increased k_La. When gas flow rate reached a certain level, the feature of mixing shock disappeared. The flow in the mixing tube was no longer a truly bubble flow, but approached an annular type of flow. Interface area per unit liquid volume was high and could not be increased by increasing gas flow rate. Therefore, k_La could not be increased any further.

Figures 4 and 5 also show that values of k_La reached above 6 s^{-1}. This is about 2 orders of magnitude higher than values obtained in stirred tanks which are typically around 0.05 s^{-1}.

4.2 Distribution of k_La

Measurements were made to determine k_La values in different sections of the Kenics mixer and ejector. The results are shown in Figure 6. For the Kenics mixer, k_La was lower in the first three and half elements of the mixer than in the last two and half. This difference may be due to an entry effect. It was visually observed that bubbles did not form until in the second element. The primary action of the first element was to divide the gas flow into two parts. Therefore, it is expected that specific interfacial area, hence k_La, in the early part of the mixer was not as high as in the later part. However the difference was small, especially at the higher flow ratio. For design purposes, it may be assumed that the mixer has an even distribution of k_La.

For the ejector, k_La was higher in the first 0.3 m of the mixing tube than in the last 0.4 m. This trend was expected in view of the presence of the mixing shock. The first section contained the mixing shock region where mixing was most intensive and therefore k_La was higher. Comparing with the Kenics mixer, this difference in k_La was much greater.

The implication of the above findings is that when designing for a Kenics mixer, the same k_La value can be used irrespective of mixer length. When designing for an ejector however, different k_La values need to be used for different mixing tube lengths (more accurately, different mixing tube length/diameter ratios).

4.3 Comparing on power basis

Power input to the two mixing devices is calculated as below.

In a motionless mixer:

$$P = Q_L \, \Delta p + Q_G \, \bar{p} \, \ln\left(\frac{p_1}{p_2}\right) + Q_L \, \frac{1}{2}\rho(u_{L1}^2 - u_{L2}^2) \tag{2}$$

In an ejector:

$$P = Q_L \, \Delta p + Q_G \, \bar{p} \, \ln\left(\frac{p_G}{p_2}\right) + Q_L \, \frac{1}{2}\rho(u_{L1}^2 - u_{L2}^2) \tag{3}$$

The last term in the above equations represents the gain in kinetic energy of the liquid phase.

For the Kenics mixer, the power comes from both liquid and gas phases; for the ejector, power comes from the liquid only, with the gas phase gaining power in most conditions (Hence the second term in equation 2 is negative).

Figure 7 shows k_La against power per unit dispersion volume at a constant gas/liquid flow ratio. Because the dispersion volumes of the Kenics mixer and ejector used in this study are very similar (Section 3), the figure is effectively a comparison of the two devices under the same power. As can be seen, k_La values at the same power per unit volume were not significantly different between the two devices.

5. INTENSIFICATION FROM A STIRRED TANK

This study has shown that k_La values in high intensity in-line mixers are much higher than in stirred tanks. what about the difference in power consumption? This section gives an exercise of intensifying a fast gas-liquid reaction from a stirred tank to an in-line mixer and compares the power and size differences between the two reactors.

Considering an intrinsically fast oxidation reaction ($A + O_2 \rightarrow B$) carried out in an existing 1.83 m diameter continuous stirred tank reactor. Assuming the reaction is carried out at 1 atm pressure and 20 ^{0}C and the tank is equipped with a Rushton turbine of standard geometry. Pure oxygen is sparged into the reactor at a rate of 0.039 m^3/s and the impeller speed is 150 rpm. Now the task is to carry out the reaction in a Kenics mixer and to achieve the same reaction rate.

Assuming the reaction falls into the moderately fast reaction regime in both Kenics mixer and stirred tank, in which case oxygen concentration in the bulk of the liquid is zero and there is no enhancement of reaction on k_L.

Based on these conditions and assumptions and using a design guide for gas-liquid mass transfer in stirred tanks (ref. 8), the oxygen transfer rate (hence reaction rate) in the stirred tank is 0.0131 kg/s.

Using another design guide for gas-liquid mass transfer in high intensity in-line mixers (ref. 9), either of the following two Kenics mixer designs can achieve the same oxygen transfer rate under a liquid flow rate of 0.02 m^3/s and gas flow rate of 0.039 m^3/s (averaged over the mixer):

1) Diameter = 0.154 m, length = 21.1 m
2) Diameter = 0.102 m, length = 4.5 m

The following table shows a comparison of reactor size and power requirement between the stirred tank and two Kenics mixer designs.

Reactor Type	Operating Mode	Size	Volume (m^3)	Power (kW)
Stirred tank	Continuous	1.83 m (dia)	4.8	3.3
Inline mixer 1 (Kenics)	Continuous	0.154 x 21.1 m (dia. x length)	0.35	9.1
Inline mixer 2 (Kenics)	Continuous	0.102 x 4.5 m (dia. x length)	0.033	23

Comparison of reactor size and power between a stirred tank and inline mixer to achieve the same reaction rate

Stirred tank: Rushton turbine, $N = 150$ *rpm,* $D = T/3$, $Q_G = 0.039\ m^3/s$
Kenics mixer: $Q_L = 0.02\ m^3/s$, $Q_G = 0.039\ m^3/s$

The table shows that the overall sizes of the in-line mixer are much smaller. The reactor volumes of the two mixer designs are respectively one or two orders of magnitude less than the stirred tank. However, the power requirements of the in-line mixer are substantially higher, about three times that of the stirred tank for design 1 and seven times for design 2.

A higher mass transfer rate (hence higher process rate) can be achieved with the in-line mixer. One way is to use the same gas and liquid flow rates and the mixer diameters shown in the table but to increase the mixer length. Obviously, the power requirement will be even higher than those shown in the table.

This comparison demonstrates that process intensification using high intensity reactors requires much higher power consumption, both to achieve the same or higher process rate. For a simple fast reaction, the main benefits from the intensification are increased process rate and improved safety (due to smaller reaction inventory). These benefits on their own may not be able to justify the large increase in power requirement. Therefore, process intensification should aim at complex reactions and give the extra potential benefit of improved product quality which arises from perhaps the effect of higher k_La and/or shorter residence time on the rates of competing reactions.

6. CONCLUSIONS

The present study has shown that volumetric mass transfer coefficient in high intensity in-line mixers are high, about 2 orders of magnitude greater that those in a stirred tank. There are similarities of mass transfer performance between motionless mixers and ejectors, e.g. in the effect of liquid flow rate and power consumption. But there are also some major differences, e.g. in the effect of gas flow rate and mass transfer in different parts of each device.

It has been quantitatively demonstrated that process intensification results in a much smaller reactor size and can increase process rate. However, the increase in power requirement is remarkable. To gain its maximum benefits and make it more acceptable, process intensification should aim at complex reactions where higher product quality can be achieved.

NOTATION

a	Specific interfacial area per unit dispersion volume	$m^2\ m^{-3}$
C_1, C_2	Oxygen concentrations in water at mixer inlet and outlet	kg/m^3
C_1^*, C_2^*	Equilibrium concentrations at mixer inlet and outlet	kg/m^3
D	Pipe or mixer diameter	m
k_L	Liquid-side mass transfer coefficient	$m\ s^{-1}$
k_La	Liquid-side volumetric mass transfer coefficient	s^{-1}
L	Length	m
Q_G	Gas flow rate averaged over a mixer	$m^3\ s^{-1}$

Q_L	Liquid flow rate	$m^3 s^{-1}$
P	Power	W
p	Pressure	Pa
Sw	Swirl number of ejector's spinner	-
u	Velocity	$m s^{-1}$
V	Gas-liquid dispersion volume	m^3

REFERENCES

1. Zhu Z. M., J. Hannon, A. Green, 1992, "Use of high intensity gas-liquid mixers as reactors". Chem. Eng. Sci., Vol. 47, pp 2847-2852.

2. Middleton J. C., 1978, "Motionless mixers as gas liquid contacting devices". A.I.Ch.E. 71st Annual meeting, Miami Beach, paper 74E.

3. Grosz-Roll F., J. Battig, and F. Moser, 1982, "Gas/liquid mass transfer with static mixing units". Proc. of 4th European Conference on Mixing, Noordwijkerhout, Netherlands, paper F2.

4. Roes A. W. M., A. J. Zeeman, and F. H. J. Bukkems, 1984, "High-intensity gas/liquid mass transfer in the bubbly flow region during co-current upflow through static mixers". I. Chem. E. Symp. Ser. No. 87, 231-238.

5. Henzler H.-J., 1980, "Model describing the suction behaviour of jet apparatus for the liquid/gaseous substance system". Chem.-Ing.-Techn., 52, 659-661.

6. Dirix C. A. M. C. and K. van de Wiele, 1990, "Mass transfer in jet loop reactors". Chem. Eng. Sci., Vol. 45, pp 2333-2340.

7. Cramers P. H. M. R., A. A. C. M. Beenackers and L. L. van Dierendonck, 1992, "Hydrodynamics and mass transfer characteristics of a loop-venturi reactor with a downflow liquid jet ejector". Chem. Eng. Sci., Vol. 47, pp 3557-3564.

8. Whitton M., 1992, "Gas-liquid mixing design guide". FMP Report 052 (confidential), BHR Group Limited, Cranfield, Bedfordshire, U.K.

9. Zhu Z. M. and M. Vernon, 1995, "Design guide for motionless mixers as gas-liquid reactors". HILINE Report HDG02 (confidential), BHR Group Limited, Cranfield, Bedfordshire, U.K.

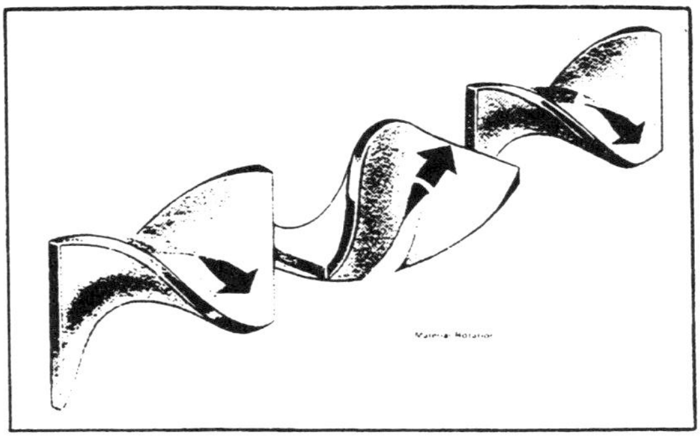

Figure 1. A Kenics mixer

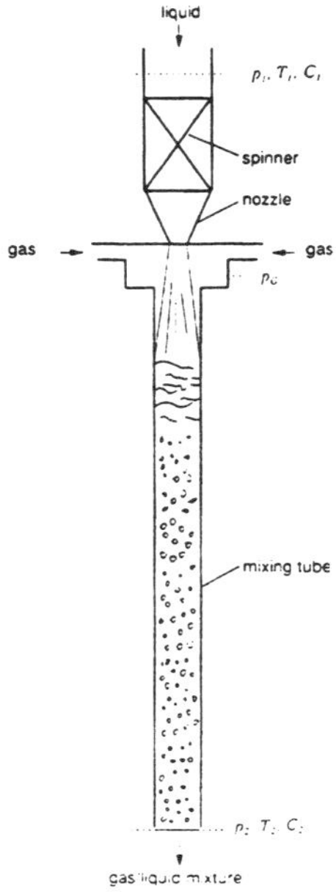

Figure 2. A gas-liquid ejector

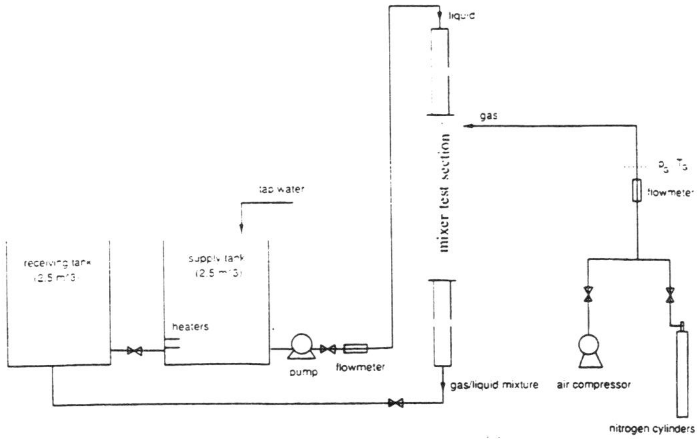

Figure 3. Experimental set-up for gas-liquid mass transfer experiments

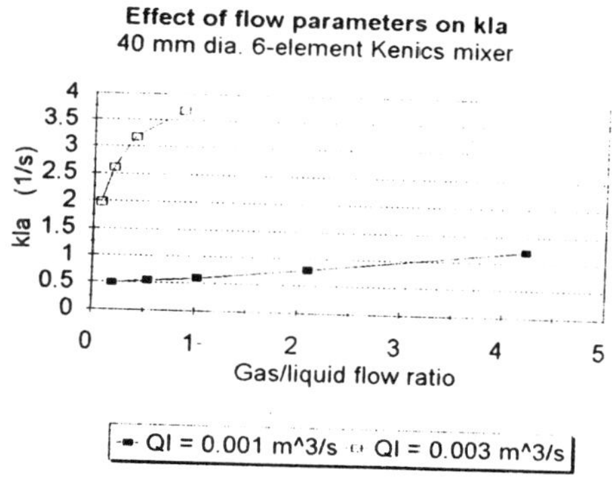

Figure 4

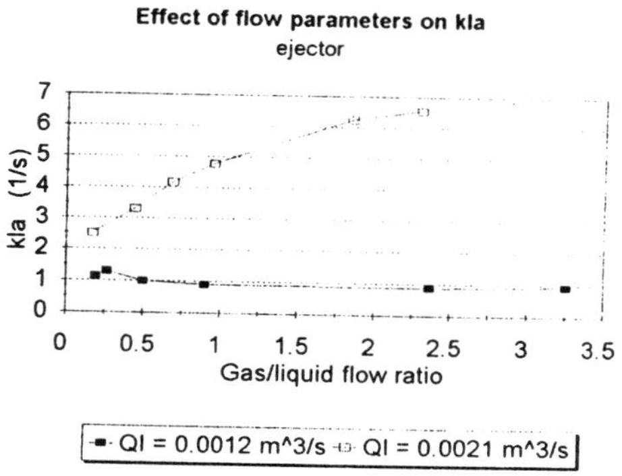

Figure 5

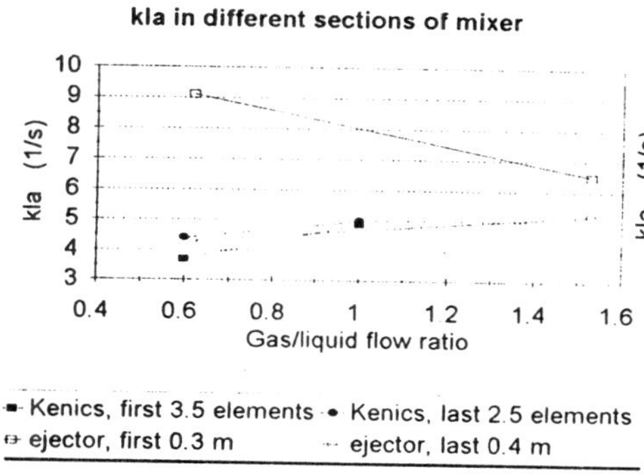

Figure 6

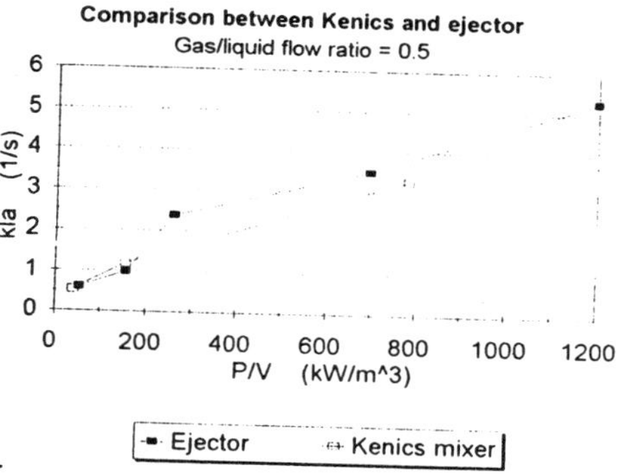

Figure 7

PULSING FLOW IN PACKED BEDS.
A METHOD OF PROCESS INTENSIFICATION.

N. A. Tsochatzidis, A. J. Karabelas *

Chemical Process Engineering Research Institute and Department of Chemical Engineering, Aristotle University of Thessaloniki, Univ. Box 455, GR 540 06 Thessaloniki, Greece.

and A. A. H. Drinkenburg

DSM, Corporate Technology, Het Overloon 1, Heerlen, P.O. Box 6500, 6401 JH Heerlen, The Netherlands.

Abstract

Pulsing (two-phase) flow is observed at relatively high gas and liquid mass flow rates and it is characterized by the regular passage of liquid-rich masses down the packed column. These pulses are essentially waves, liquid-continuous in character, picking up liquid at the front, and after mixing, leaving it at the tail. The wave velocity is usually smaller than the mean gas velocity, so that gas "penetrates" the pulse from behind (in the form of bubbles) and escapes from the front. This intensive interaction between the phases is very beneficial for many chemical and physical processes. Potential advantages of pulsing flow include *capacity increase*, mainly through the enhanced (gas-liquid and liquid-solid) mass and heat transfer rates, as well as *elimination of hot spots* in fast exothermic reactions, and *reduced axial mixing* due to the "washing" action of the liquid-rich pulses.

The purpose of this paper is to outline research results, obtained by the authors in two different laboratories, which are helpful in describing pulsing flow and in understanding the dominant mechanisms. Information is summarized on the basic pulse characteristics (frequency, celerity, length, duration), liquid holdup, pressure drop, as well as gas-liquid and liquid-solid mass transfer rates. On the basis of this information research directions are identified which would facilitate industrial applications.

* Author for correspondence.

INTRODUCTION

There are significant industrial applications of packed beds with cocurrent downflow of gas and liquid (often called *trickle beds*); e.g. Dudukovic & Mills (1986). Four main flow regimes have been identified in operation of this equipment, i.e. *trickling*, *pulsing*, *spray* and *bubble* flow regimes, as shown in the simplified diagram of Figure 1. At present, "trickling" is the most common regime in industrial applications. This regime has well-known advantages and limitations discussed in several review papers that have appeared in the literature over the past twenty years (e.g. Satterfield, 1975; Charpentier, 1976; Hofmann, 1977; Herskowitz & Smith, 1983; Koros, 1986; Gianetto & Specchia, 1992).

This paper is focused on *pulsing* flow, stressing the potential advantages of this mode of operation as a method of process intensification. The present review is based on research results obtained by the authors in two different laboratories; they include data obtained by Drinkenburg and associates (Blok & Drinkenburg, 1982; Blok et al., 1983; Rao & Drinkenburg, 1983; Rao & Drinkenburg, 1985b) as well as the more recent results reported by Tsochatzidis & Karabelas (1994a, 1994b, 1995a). Information from all the above papers, combined, is perhaps the most complete set of data on pulse characteristics (for fixed liquid and gas physical properties) available in the literature at present. Only air and water, or liquids of similar physical properties, are employed in these studies.

In this paper the macroscopic hydrodynamic characteristics of pulsing flow (pulse frequency, celerity, length, duration, pressure drop, liquid holdup) are summarized first and some general observations are made. Typical data of mass transfer rates in pulsing flow provide convincing evidence of transport process intensification. Next, the results of information extracted at the microscopic scale (using fast video recordings and localized probes embedded in the packing) are outlined, which are essential for interpreting the macroscopic measurements. Following that, a brief physical description of pulsing flow is provided, which is helpful for modeling, for scale-up considerations and for understanding the trends of measured transport coefficients. Finally, comments are made about the applicability of the present results to industrial processes and on R&D needs to extend the range of system variables tested and to increase confidence in scaling-up to industrial size beds.

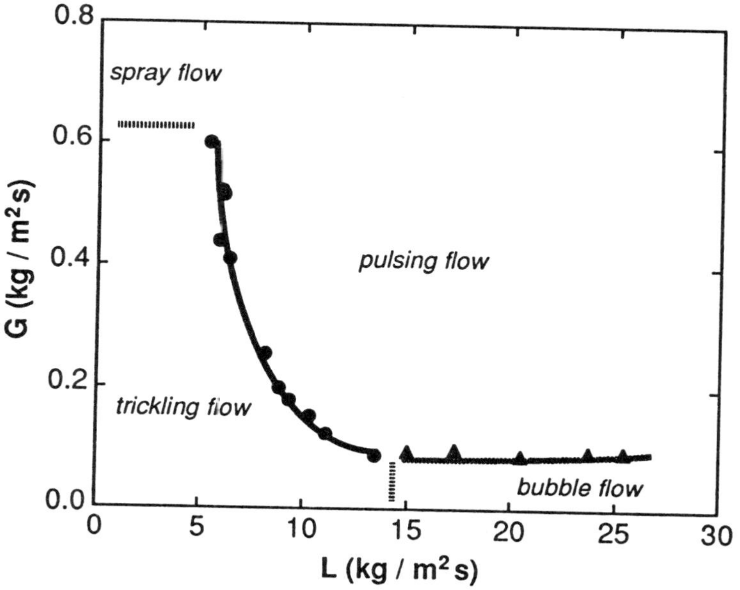

Figure 1. Simplified flow map of a trickle bed. G and L represent air and water mass fluxes, respectively. Adapted from Tsochatzidis & Karabelas (1994a).

MACROSCOPIC PULSING FLOW CHARACTERISTICS

Pulsing flow is usually described by means of its *macroscopic* hydrodynamic properties, which are readily observable and measurable over a sufficiently long time period so that the entire process is at steady state. Most of these properties are of practical interest in designing packed beds as well as in developing meaningful correlations of transport coefficients.

General Description

For recording and analyzing pulsing flow, a conductance technique was employed by Tsochatzidis & Karabelas (1995a) which is capable of providing accurate instantaneous measurements of pulsing flow characteristics. Two identical ring probes

(Tsochatzidis et al., 1992) were installed in the lower half of the packed bed, providing a satisfactory representation of the conditions in a cross-section of the bed, with sharp definition of pulses. An example of two simultaneously recorded signals is shown in Figure 2 (traces a,b). It is evident that each peak corresponds to the passage of the liquid-rich zone of a pulse, followed by the low conductance (gas-rich) zone. The two traces are nearly identical except for a time-lag, corresponding to the time required by each pulse to cover the distance between the probes (20 cm). A steep increase is observed in the signal (corresponding to the liquid holdup) associated with the front of the liquid-rich zone followed by a more gradual decrease in the gas-rich zone, which makes difficult the precise determination of boundaries between the liquid- and the gas-rich zones. Occasionally double or (seldom) triple pulses are produced, as shown in Figure 2. As the gas and liquid flow rates increase, the generation of double or triple pulses tends to increase. Statistical analysis of time records from the conductance probes was performed to determine the following main pulse characteristics.

Trace c of Figure 2 is from a pressure transducer placed at the same bed axial location as the lower ring probe (trace b). It is apparent that there is a time lag between the peaks of traces b and c. The pressure peaks lag behind the liquid holdup maxima. For the conditions of these experiments (Tsochatzidis & Karabelas, 1995a), the corresponding mean distance between peaks is in the range 4.0 - 7.5 cm. The pressure peak corresponds to the back of the liquid-rich zone and essentially provides the force required to push the liquid-rich zone downstream.

The *onset of pulsing* is quite predictable on the basis of available experimental data. By increasing the flow rates of a bed operating under trickling flow conditions, pulses are detected first at the bottom of the bed, where the gas velocity is comparatively higher. A further small increase in the flow rates beyond the transition, raises sharply the pulsing onset level in the column. In fully developed pulsing flow, pulses are formed approx. 10 cm below the column entrance.

Pulse Properties

Values of *pulse frequency* on the order of 10 Hz are reported for various flow rates, bed and packing characteristics. The pulse frequency tends to increase with both gas and liquid superficial velocities. However, for high values of liquid velocity, it seems to depend only on the gas flow rate as reported by Tsochatzidis & Karabelas (1995a). Generally there is some uncertainty in the precise determination of the pulse characteristic frequency mainly due to the coalescence of pulses as they travel down the bed and to the existence of double (or occasionally triple) pulses.

Pulse celerity is of order 1 m/s for the systems studied (Rao & Drinkenburg, 1985a). For small liquid flow rates, pulse celerity depends on the gas flow rate while a very weak influence (or no influence) of the liquid flow rate is reported. For higher

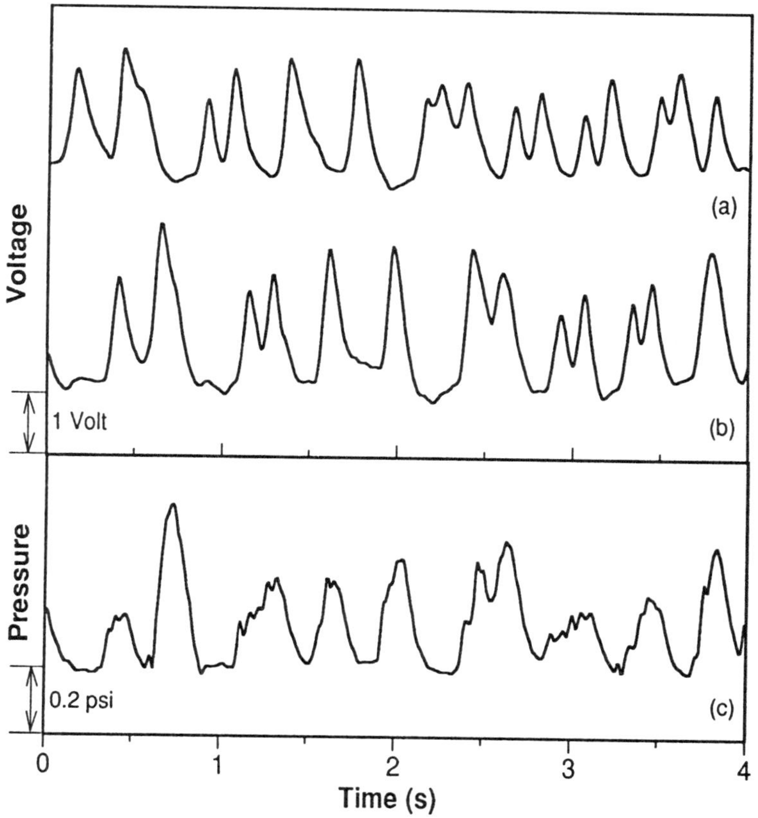

Figure 2. Typical simultaneous traces in the pulsing flow regime from conductance ring probes (a, b) and a pressure transducer (c). Flow rates: L=20.509 kg/m^{2}s, G=0.257 kg/m^{2}s. Adapted from Tsochatzidis & Karabelas (1995a).

liquid rates the influence of liquid flow tends to increase as shown in Figure 3. At high real gas velocities, the pulse celerity data tend to approach an asymptotic value.

Over a large range of gas flow rates, the pulse celerity is *smaller* than the real gas velocity. In these cases, gas must *penetrate* the pulse from behind and gas bubbles or fingers must travel through the liquid-rich zone of the pulse. As outlined below, visual observations (Tsochatzidis & Karabelas, 1994a) confirm this result. The value of gas velocity at which this penetration mechanism starts is expected to be a function of the liquid surface tension according to Blok & Drinkenburg (1982).

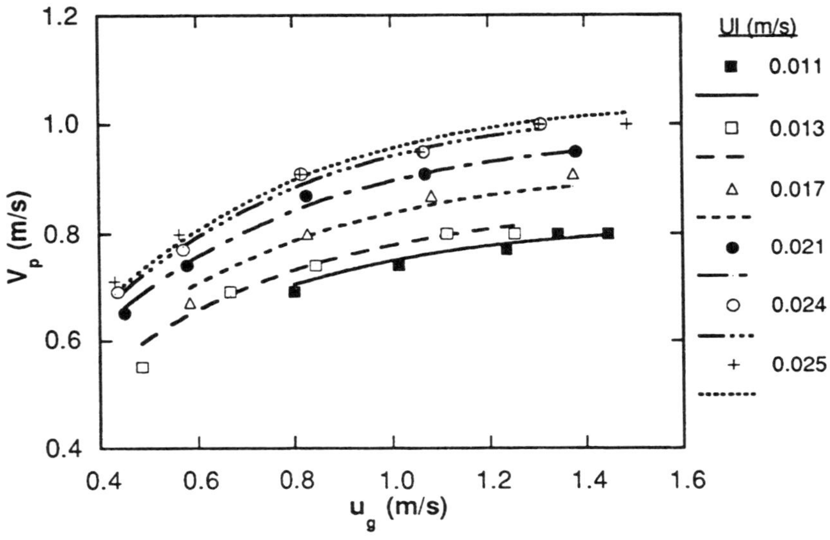

Figure 3. Influence of real gas velocity, u_g, and of superficial liquid velocity, U_l, on pulse celerity, V_p. Adapted from Tsochatzidis & Karabelas (1995a).

The *pulse length* includes both the liquid-rich and gas-rich zones of a pulse. For high liquid flow rates, the pulse length seems to be almost constant and roughly equal to 0.3 m. Detailed data on the length of moving liquid-rich zones are reported by Tsochatzidis & Karabelas (1995a). These data show that there is a distribution of liquid-rich lengths, with typical values (over a rather broad range of flow rates) close to 10 cm. The gas-rich zone is normally larger than the respective liquid-rich zone of a pulse.

The *duration* of the liquid- or of the gas-rich zone of pulses is expected to influence the bed pressure drop and the overall mass and heat transfer rates. The gas-rich (or equivalently the liquid-rich) duration seems to depend only on the liquid and not on the gas flow rate (Tsochatzidis & Karabelas, 1995a). This may imply that the liquid input rate is mainly responsible for the pulse geometry.

It will be noted that the influence of the column diameter on pulse properties is insignificant (Blok & Drinkenburg, 1982), for relatively small column I.D; i.e. 5 to 20 cm. It is, however, impossible at present to extrapolate this result to much larger diameter columns.

Two-phase pressure drop

The two-phase pressure drop per unit length of the bed tends to increase with both gas and liquid superficial velocities in the pulsing flow regime. This increase is almost linear (Rao & Drinkenburg, 1983; Tsochatzidis & Karabelas, 1995a). An example of the pressure drop variation for 3 mm spheres is depicted in Figure 4 in which the arrows mark the transition from trickling to pulsing flow regime. It must be pointed out that no sharp change in the pressure drop is observed at the transition; i.e. the pressure drop does not appear to be greatly influenced by pulsing. A model for pressure drop is presented by Rao & Drinkenburg (1985a) in which the two-phase pressure drop in the pulsing flow regime is the sum of contributions from the gas continuous part (outside the pulses) and from the pulses themselves. Other pressure drop correlations for pulsing flow can be found in Tsochatzidis & Karabelas (1995a) and Rao & Drinkenburg (1983).

Liquid holdup

The liquid holdup is the ratio of the liquid volume to the total volume of the bed and is comprised of the dynamic and the static component (for non porous packing material). Figure 5 shows that the mean dynamic liquid holdup tends to *decrease* with increasing gas flow rate. A relatively small increase of the mean liquid holdup is observed by increasing the liquid flow rate. Blok et al. (1983) also report that at a certain gas flow rate (beyond the onset of pulsing) the liquid holdup becomes practically constant. Higher liquid flow rates are obviously accommodated mainly by an increase of pulse celerity and frequency and to a limited extent through an increase of liquid holdup.

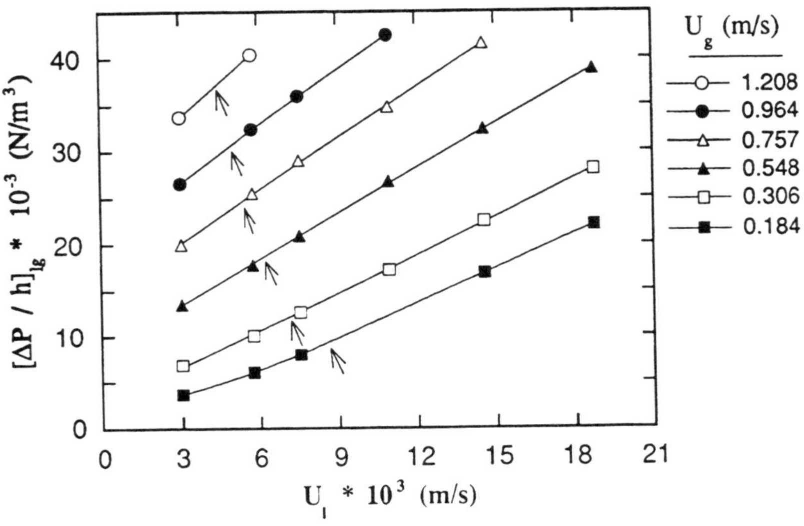

Figure 4. Variation of two-phase pressure drop in a trickle bed of 3 mm spheres, with gas and liquid superficial velocities. Adapted from Rao & Drinkenburg (1983).

The pulsing flow regime is characterized by a significant temporal variation of the liquid holdup. Peak values approx. 40% over the mean holdup are associated with the liquid-rich pulses. On the other hand gas-rich zones (of relatively longer duration) do not exhibit a similar holdup reduction (Tsochatzidis & Karabelas, 1995a). The dynamic liquid holdup in the liquid-rich, $h_{d,l-r}$, and in the gas-rich, $h_{d,g-r}$, zone of a pulse have also been determined. In general, the influence of gas and liquid flow rates on $h_{d,l-r}$ and $h_{d,g-r}$ is similar to that observed for the mean liquid holdup (Tsochatzidis, 1994). Data of $h_{t,l-r}$, $h_{t,g-r}$ and of the mean total liquid holdup, h_t, reported by Tsochatzidis & Karabelas (1995a) for 6 mm spheres are in fair agreement with correlations of these quantities proposed by Blok & Drinkenburg (1982) and Blok et al. (1983) for 2.5x2.5 mm and 4x4 mm Raschig rings and column I.D. 5, 10 and 20 cm.

It should be noted that in the data of Tsochatzidis & Karabelas (1995a) the ratio $h_{t,l-r} / h_{t,g-r}$ is almost constant ranging from 1.2 to 1.5 for the flow rates examined. A

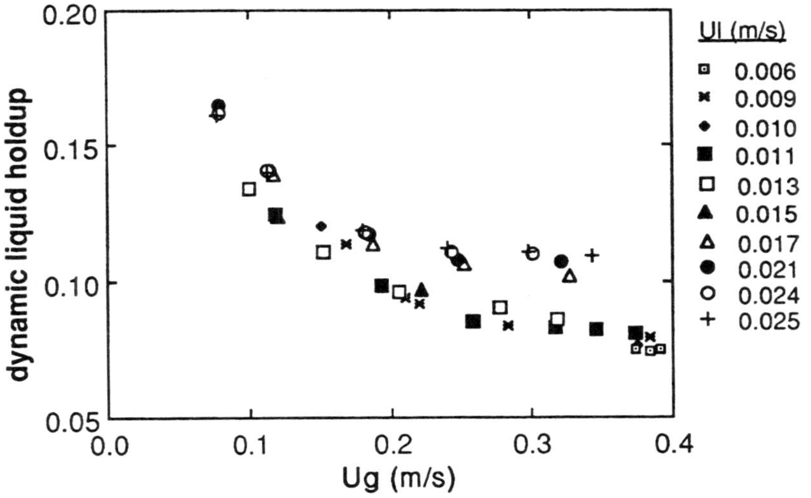

Figure 5. Influence of gas and liquid superficial velocities on the dynamic liquid holdup in the pulsing flow regime. Adapted from Tsochatzidis & Karabelas (1995a).

somewhat larger range (1.5 to 2) is observed by Blok & Drinkenburg (1982). For high flow rates this ratio could be taken as constant and equal to 1.5.

For their column, Tsochatzidis & Karabelas (1995a) report that the highest measured total holdup value in the liquid-rich zone is 0.22, compared to a maximum of 0.36 for perfect flooding. This leads to the conclusion that the liquid-rich zone of a pulse "contains" a large amount of gas in the form of bubbles or clusters. Visual observations confirm this conclusion (Tsochatzidis & Karabelas, 1994a). The highest liquid fraction is measured at the front of the liquid-rich zone. The presence of dispersed bubbles is also observed in this zone by Blok & Drinkenburg (1982).

TRANSPORT COEFFICIENTS

Data on transport rates provide strong evidence that pulsing flow is an effective means of intensifying physico-chemical processes taking place in packed beds.

Gas-liquid mass transfer

Gas-liquid mass transfer rates in the pulsing flow regime are much larger than those of the trickling regime and larger than those obtainable in countercurrent flow in beds of similar properties (Seirafi & Smith, 1980; Gianetto et al., 1973). A sharp increase in the gas-liquid mass transfer rate, ak_L, is observed in the transition region from trickling to pulsing flow as depicted in Figure 6 where data from Seirafi & Smith (1980) are presented. Transfer rates more than 100% higher than in trickling flow are reported for the pulsing flow regime. This increase is attributed to the large interfacial area of the small bubbles dispersed in the liquid-rich zone (e.g. Hirose et al., 1974) and to the intensive gas-liquid interactions. Evidence in support of this assertion, as well as a more precise physical picture, are given below.

Solid-liquid mass transfer

In Figure 7 the solid-liquid mass transfer rates (in terms of the wetted fraction of particles, φ, times the Sherwood number, Sh) are presented for the trickling, transition and the pulsing flow regime. The measured transfer rates in the gas-rich zone (base), the liquid-rich zone (pulse) and their average value are included. Data are taken with an electrodiffusion method using one active particle placed at the center of the bed comprised of 6 mm cylinders (Rao & Drinkenburg, 1985b). A rather steep increase is observed in the transition region, which is also reported by Tsochatzidis & Karabelas (1995b) for 6 mm spheres. The liquid-rich mass transfer rate is about 2.5 times greater than that corresponding to the gas-rich zone. A relatively small radial variation of the time averaged mass transfer coefficient, k_S, is found by Tsochatzidis & Karabelas (1994b). Lower values of φk_S are obtained close to the bed wall while throughout the rest of the bed the variations appear to be insignificant.

Further interpretation of the mass transfer measurements leads to the following results (Tsochatzidis & Karabelas, 1994b). In the liquid-rich zone apparently complete and radially uniform wetting takes place. In the gas-rich zone considerably smaller mass transfer rates prevail. Local voidage nonuniformities are more evident (through the mass transfer data) in this part of the pulse. The time averaged mass transfer rates over a particle are mainly influenced by the high values associated with the liquid-rich zone of the pulse.

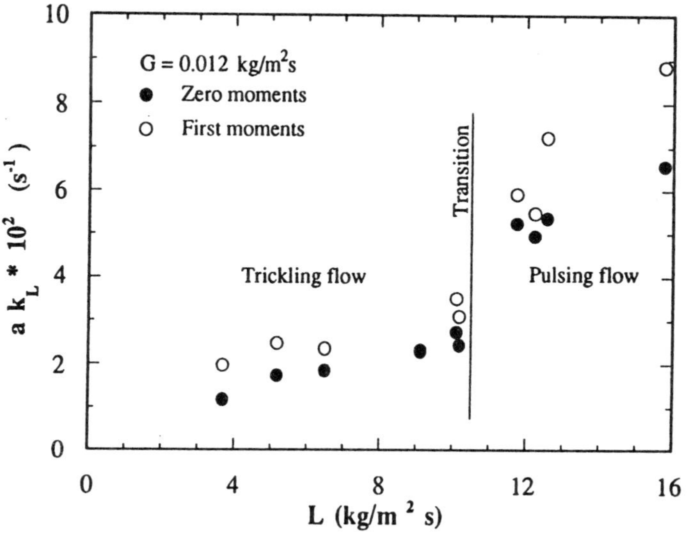

Figure 6. Variation of gas-liquid mass transfer coefficient in the trickling and pulsing flow regime. Adapted from Seirafi & Smith (1980).

MICROSCOPIC CHARACTERISTICS

The above macroscopic properties are actually manifestations of elementary processes occurring at a much smaller length scale, and over a much shorter time scale (Tsochatzidis & Karabelas, 1994a). To amplify this statement one should recognize that the complicated flow channels in a packed bed are a network of irregular *pores* interconnected through several narrow passages or *constrictions*. As discussed here, these geometrical features play a very important role in flow pattern development.

Visual observations

These were made using a fast video camera (Tsochatzidis & Karabelas, 1994a). A *gas penetration* mechanism appears to be the dominant feature of pulsing flow in the microscale. In the liquid-rich zone, bubbles tend to form rapidly at the constrictions of the packing material, covering the available space in a pore and vanishing over a time period of order 10 ms as clearly depicted in the sequence of pictures of Figure 8. These

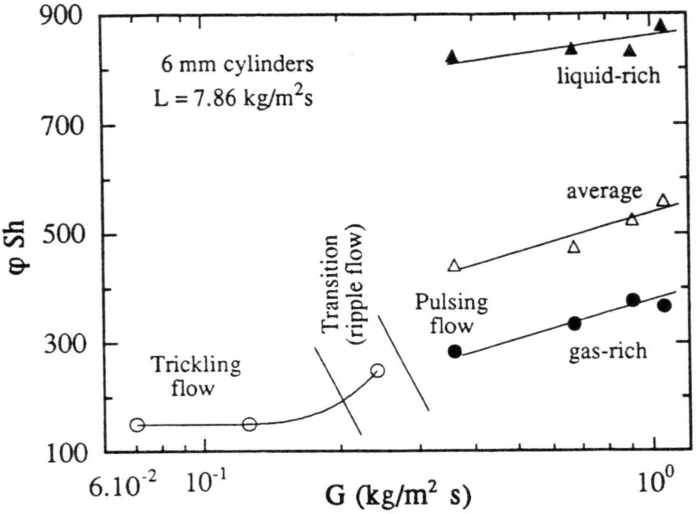

Figure 7. Comparison of solid-liquid mass transfer rates in the trickling and pulsing flow regime. Adapted from Rao & Drinkenburg (1985b).

pictures show typical flow conditions in the packing interstices during the passage of a liquid-rich zone of the pulse. This small characteristic time period is due to the relatively high real gas velocity (of order 1 m/s) and to the small (a few millimeters) characteristic pore size. It should be noted that the time scale of these events is *much shorter* than the corresponding time macroscale which can be determined using the pulse frequency.

This gas penetration mechanism is considered very significant for intensifying transport processes, causing high gas-liquid interface renewal rates, increasing convection of the liquid film covering the particles, and promoting liquid mixing. Due to this rapid bubble generation and penetration through the constrictions into the (randomly arranged) pores, there is no instance during the passage of the liquid-rich zone that pores on a bed cross-section are totally liquid-filled. This is also concluded from the measured liquid holdup, which is always much below its flooding value.

Rao & Drinkenburg (1985a), reporting on their visual observations, consider the liquid-rich zones similar to dispersed bubble flow through a packed bed. However, the

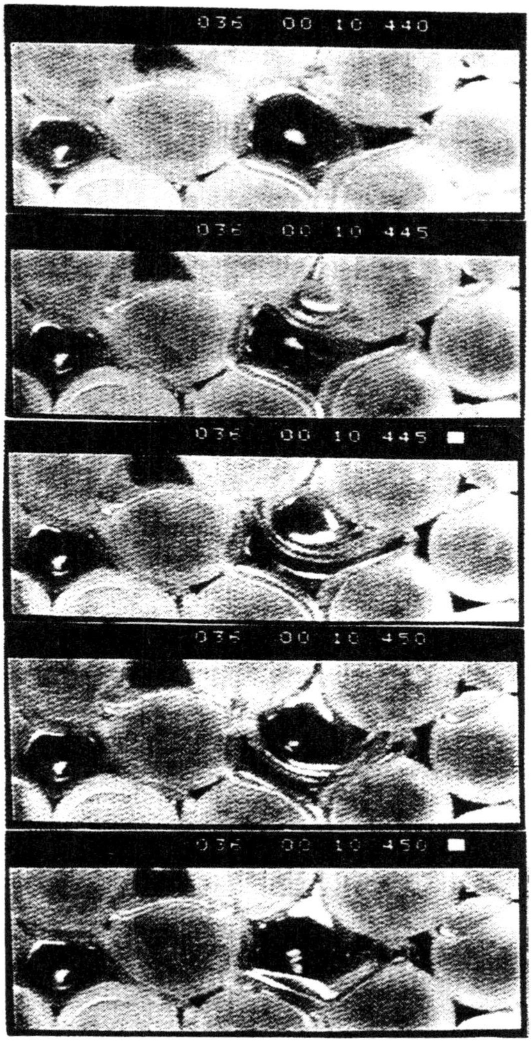

Figure 8. Mode of gas penetration into pores during the passage of the liquid-rich zone of a pulse. A bubble is developing into the large depicted pore. Adapted from Tsochatzidis & Karabelas (1994a).

gas-rich zone of a pulse appears to be quite similar to trickling flow, with liquid wetting the particles in a gas-continuous flow field.

Local Measurements

For flow diagnostics in the microscale, electrodes of the same dimensions as the packing material and microelectrodes embedded onto the surface of a particle are employed by Tsochatzidis & Karabelas (1994b) with an electrodiffusion technique. Measurements show that pulses cover the entire bed cross section and the pulse disturbances nearly uniformly penetrate even the smallest flow channels, resulting in periodic fluid renewal therein. Furthermore, complete and radially uniform wetting appears to prevail during pulsing flow. Independent measurements with a local conductance probe employed by Tsochatzidis & Karabelas (1995a), in order to examine flow conditions in the core of the bed, support the above conclusions. It must be pointed out that the above results (complete wetting, radial uniformity) may not apply to much larger bed diameters.

BRIEF PHYSICAL DESCRIPTION

Macroscopically pulsing flow appears to be a succession of zones. A zone resembling moving bubble flow (the liquid-rich zone of a pulse) and another similar to trickling flow (the gas-rich zone). In the latter rivulets are observed, as in the trickling regime, while in the liquid-rich zone the bubble activity seems to be greater than in the regular bubble flow pattern. In general, the pulses appear to be axisymmetric with their front on a horizontal plane (Tsochatzidis & Karabelas, 1995a). For relatively small bed diameters, (< 30 cm) pulses apparently span the entire column cross-section. Pulses not spanning the cross-section, as reported by Christensen et al. (1986) for their two-dimensional packed bed might characterize columns of larger diameters.

A fairly good quantitative description of main pulse properties (celerity, frequency, length, duration) is available. In general, it appears that the liquid flow rate tends to influence more the "geometrical" features of the pulse (length, duration), while the gas throughput provides the driving force. It is important to note that even the liquid-rich zone of a pulse is associated with a total holdup far from the flooding condition of the bed. Thus, pulses are made up of a two-phase mixture. Even when real gas and liquid velocities in the pulse are roughly the same, the passage of this two-phase mixture through the narrow constrictions will improve interphase contact and increase liquid mixing.

However, the processes occurring in the packing interstices appear to be greatly intensified by the "gas penetration" mechanism which was found to take place over most of the domain of pulsing flow. In essence, on the motion of a liquid-rich pulse there is another process (gas penetration) superposed, of a *much shorter* time scale, leading to repeated rapid formation and vanishing of gas bubbles. This mechanism appears to be responsible for large interface renewal rates (i.e. large gas/liquid mass transfer), increased liquid film convective velocities (i.e. large solid/liquid mass transfer) and improved liquid mixing, expected to have a beneficial effect in chemical reactions. It is somewhat surprising that this intensification is achieved with no penalties in pressure drop, since transition to pulsing flow is not accompanied by unusual pressure losses.

DISCUSSION

Among the distinct advantages of pulsing flow is the intensive interaction between the phases due to the gas penetration mechanism, elimination of hot spots due to the frequent passage of the pulses, and reduced axial mixing (reduced backmixing). High mass transfer rates are observed, while the selectivity of a process can be increased. This *self-induced* pulsing flow can benefit several industrial processes. For example, this flow regime is well suited for exothermic reactions with fast kinetics. However, further study is clearly needed to extend the range of system variables tested and to improve confidence for possible industrial applications. The effect of liquid viscosity and surface tension on pulsing flow characteristics and on the onset of pulsing is worth studying. The influence of large bed diameters on pulse uniformity and liquid distribution is also a major unresolved question. Additionally, modeling efforts of predicting pulsing flow characteristics are certainly required. A model of this type is presented by Tsochatzidis & Karabelas (1995b) and allows one to estimate pulse characteristics as well as average quantities such as total pressure drop and liquid-solid mass transfer coefficients. A pressure drop model for the pulsing flow regime is also presented by Rao & Drinkenburg (1985a). Reliable models are obviously still in demand for scale-up to industrial sizes.

Restrictions of pulsing flow seem to arise in the case of operation at elevated pressures where the gas density is increased, the liquid holdup is decreased and consequently higher liquid feed is necessary for the onset of pulsing; thus, the region of trickling flow becomes larger (Wammes et al., 1990).

Conditions similar to self-induced pulsing flow may be achieved by *periodic operation* of a trickle bed. In this case the gas phase is introduced into the packed bed continuously while the liquid phase is usually fed in an on-off mode. Haure et al. (1989) report an increase in the oxidation rate of SO_2 by about 30 to 45% in a periodically operated trickle bed, compared to the steady operating conditions. Improved time average conversion (and selectivity) relative to that under steady-state operation is also reported by Lange et al. (1994) which is attributed to increased catalyst wetting, the development of more liquid channels in the bed, improved utilization of liquid holdup and temperature oscillations. All of the above are characteristics associated with pulsing flow. It is also reported that liquid cycling has a beneficial influence on the conversion. The above results suggest that the *induced pulsing flow* (essentially a kind of periodic operation of the trickle bed) deserves further study.

Acknowledgment

The authors from CPERI, Thessaloniki, are grateful to the Commission of the European Communities for financial support of their work under contracts JOUG-0005-C and JOU2-CT92-0067.

References

Blok, J. R. and Drinkenburg, A. A. H., 1982, Hydrodynamic properties of pulses in two-phase downflow operated packed columns. *Chem. Eng. J.* **25**, 89-99.

Blok, J. R., Varkevisser, J. and Drinkenburg, A. A. H., 1983, Transition to pulsing flow, holdup and pressure drop in packed columns with cocurrent gas-liquid downflow. *Chem. Eng. Sci.* **38**, 687-699.

Charpentier, J. C., 1976, Recent progress in two phase gas-liquid mass transfer in packed beds. *Chem. Eng. J.* **11**, 161-181.

Christensen, G., McGovern, S. J. and Sundaresan, S., 1986, Cocurrent downflow of air and water in a two-dimensional packed column. *AIChE J.* **32**, 1677-1689.

Dudukovic, M. P. and Mills, P. L., 1986, Contacting and hydrodynamics in trickle-bed reactors. *Encyclopedia of Fluid Mechanics* (Edited by N.P. Cheremisinoff), Gulf Publishing Company, Vol. **3**, pp. 969-1017.

Gianetto, A. and Specchia, V., 1992, Trickle-bed reactors: State of art and perspectives. *Chem. Eng. Sci.* **47**, 3197-3213.

Gianetto, A., Specchia, V. and Baldi, G., 1973, Absorption in packed towers with concurrent downward high velocity flows - II: Mass transfer. *AIChE J.* **19**, 916-922.

Haure, P. M., Hudgins, R. R. and Silveston, P. L., 1989, Periodic operation of a trickle-bed reactor. *AIChE J.* **35**, 1437-1444.

Herskowitz, M. and Smith, J. M., 1983, Trickle-bed reactors: A review. *AIChE J.* **29**, 1-18.

Hirose, T., Toda, M. and Sato, Y., 1974, Liquid phase mass transfer in packed bed reactor with cocurrent gas-liquid downflow. *J. Chem. Eng. Jpn.* **7**, 187-192.

Hofmann, H., 1977, Hydrodynamics, transport phenomena, and mathematical models in trickle-bed reactors. *Int. Chem. Eng.* **17**, 19-28.

Koros, R. M., 1986, Engineering aspects of trickle bed reactors, in *Chemical Reactor Design & Technology* (Edited by H. I. de Lasa), NATO ASI Ser., Series E: Appl. Sci. No 110, M. Nijhoff Publishers, Dordrecht, pp. 579-630.

Lange, R., Hanika, J., Stradiotto, D., Hudgins, R. R. and Silveston, P. L., 1994, Investigations of periodically operated trickle-bed reactors. *Chem. Eng. Sci.* **49**, 5615-5621.

Rao, V. G. and Drinkenburg, A. A. H., 1983, Pressure drop and hydrodynamic properties of pulses in two-phase gas-liquid downflow through packed columns. *Can. J. Chem. Eng.* **61**, 158-167

Rao, V. G. and Drinkenburg, A. A. H., 1985a, A model for pressure drop in two-phase gas-liquid downflow through packed columns. *AIChE J.* **31**, 1010-1018.

Rao, V. G. and Drinkenburg, A. A. H., 1985b, Solid-liquid mass transfer in packed beds with cocurrent gas-liquid downflow. *AIChE J.* **31**, 1059-1068.

Satterfield, C. N., 1975, Trickle-bed reactors. *AIChE J.* **21**, 209-228.

Seirafi, H. A. and Smith, J. M., 1980, Mass transfer and adsorption in liquid full and trickle beds. *AIChE J.* **26**, 711-717.

Tsochatzidis, N. A. and Karabelas, A. J., 1994a, Experiments in trickle beds at the micro- and macroscale. Flow characterization and onset of pulsing. *Ind. Eng. Chem. Res.* **33**, 1299-1309.

Tsochatzidis, N. A. and Karabelas, A. J., 1994b, Study of pulsing flow in a trickle bed using the electrodiffusion technique. *J. Appl. Electrochem.* **24**, 670-675.

Tsochatzidis, N. A. and Karabelas, A. J., 1995a, Properties of pulsing flow in a trickle bed. *AIChE J.* (in press).

Tsochatzidis, N. A. and Karabelas, A. J., 1995b, Characteristics of pulsing flow in trickle beds. Improvements of a macroscopic model. *Chem. Eng. Commun.* **137**, 85-100.

Tsochatzidis, N. A., 1994, A study of two-phase cocurrent downflow in packed beds: Pulsing flow. Doctoral Dissertation, Dept. of Chemical Engineering, Aristotle University of Thessaloniki, Greece.

Tsochatzidis, N. A., Karapantsios, T. D., Kostoglou, M. V. and Karabelas, A. J., 1992, A conductance probe for measuring liquid fraction in pipes and packed beds. *Int. J. Multiphase Flow* **18**, 653-667.

Wammes, W. J. A., Mechielsen, S. J. and Westerterp, K. R., 1990, The transition between trickle flow and pulse flow in a cocurrent gas-liquid trickle-bed reactor at elevated pressures. *Chem. Eng. Sci.* **45**, 3149-3158.

SYSTEMS

1st International Conference on:
Process Intensification for the Chemical Industry
Antwerp, 6-8 Dec. 1995

Potential Benefits of Using Compact Multi-stream Heat Exchangers in Integrated Process Plants

J.van Reisen(*), P.J.T.Verheijen(*) & G.T.Polley(+)
* Dept. of Chemical Engineering, TU Delft (NL)
+ Dept. of Chemical Engineering, UMIST (UK)

Multi-stream heat exchangers have been used in cryogenic applications for a number of years. Interest has recently switched to their use in general petro-chemical applications(1). In this paper we review the advantages and disadvantages of multi-stream exchangers. We then consider their potential application in both new plant design and energy saving revamps.

Multi-stream Exchangers

Multi-stream exchangers are defined as those in which three or more streams are thermally processed in a single unit. Although rarely applied, these units offer significant advantages in integrated plants. They can be used for conventional heat recovery between a multitude of streams or, they can be used to make heat recovery possible. For instance, a hot stream that is just a few degrees higher than a reboiler stream can be used once boiling has commenced but is not itself able to provide the wall superheat necessary for initial nucleation. Under these circumstances, hot utility could be used in one part of an exchanger to initiate boiling, whilst the hot stream is subsequently used in another section to provide the bulk of the boil-up.

There are currently two principal types of heat exchangers that can be used as multi-stream units. These are the plate-fin exchanger and the plate-and-frame exchanger. Wound coil units are also available for use in specialised situations.

Multi-stream plate-fin exchangers have been used in cryogenic plants for a number of years. Given their proven thermodynamic efficiency (see below) it is not surprising that their use in integrated processes is now attracting attention.

Plate-and-frame exchangers have found multi-stream use in the dairy industry. However, application of multi-streaming in other process industries appears to be rare. This is a pity for their use as multi-stream units is not particularly difficult to achieve. Conventional plates each have four ports. Likewise, the headers each have four ports. So, each exchanger has eight ports. Since, any stream just requires an inlet and an outlet port, a conventional unit can handle two, three or four streams. All that is required is the installation of an intermediate plate on which the ports have not been cut. Possible three and four stream arrangements are shown in Figure 1.

With the introduction of special intermediate plates yet more streams can be handled. For instance, the configuration of a five stream unit is illustrated in Figure 2.

Advantages and Disadvantages of Multi-streaming

Figure 3 shows a set of Composite Curves typical of a crude oil pre-heat train. (Suitable units do not yet exist for this application but the problem can be used for illustration). If a high energy efficiency is to be achieved the designer must ensure two things. First, there has to be a close temperature approach at the 'Pinch' (the practical limit for shell-and-tube exchangers is around 10C). Second, the hot and cold streams that occur in the same 'enthalpy interval' should be matched for heat recovery. The matching of hot and cold streams in this way often leads to stream splitting. It is not surprising to find that although the introduction of 'Pinch Technology' simplified the process of designing heat recovery networks the results were structures of greater complexity. In Figure 4 we show a crude oil pre-heat train that was designed before the advent of Pinch Technology. In Figure 5 the design developed using Pinch Technology is shown. There is a marked increase in both the number of stream splits and number of exchangers used.

In the above example it is clear that high energy efficiency is often purchased by higher capital expenditure both in terms of higher exchanger cost and higher piping costs. Multi-streaming offers a way around both of these costs. Since, more than one duty can be accomplished in a single unit, the number of exchangers that have to be purchased is reduced. A particularly significant saving in installation costs can be expected. Since, stream splitting is undertaken inside multi-stream units (through the internal plate and channel arrangments) plant piping will be greatly simplified and significant cost savings made.

Multi-stream exchangers also tend to be far more 'compact' than conventional exchangers. This leads to further savings in installation costs. More significantly, it can directly lead to reductions in plant plot size and layout associated costs.

Finally, compact exchangers generally employ pure counter current flow (as opposed to the repetative crossflow used in shell-and-tube heat exchangers). This means that they can be used a much closer temperature approaches (down to 1C if necessary). In some processes, typically cryogenic ones, this is a very important property leading to significant energy savings.

The principle disadvantage is that multi-streaming is not developed as a general technology. Whilst it may be used in some 'special' processes (e.g. cryogenic ones) it has not been employed in the process industries in general. User experience does not exist. Design experience does not exist.

There are concerns regarding the operabilty of multi-stream units. The interactions between the streams can be complex (see Figure 6). So, how easy are they to control?

Efficient operation of such a unit requires that the streams matched against a common stream (i.e. the hot streams matched against a specific cold stream across the same temperature span) have closely matched temperatures. So, how flexibile are such units? Do they stil operate efficiently if the flowrate of one of the streams changes?

All of these questions are currently 'research questions'. Before investment in such research can be justified the potential benefits of multi-streaming must be demonstrated. The areas that such technology is best employed must be established.

Application in New Plant Design

In terms of heat network design, crude pre-heat trains present a special case, for there is generally just one cold stream (the crude oil) but several available hot streams. This cold stream must also be considered to consist of three separate sections: that prior to the desalter, that between the desalter and the flash column, and that between the flash column and the fractionation tower.

Since, there is just one cold stream, it can be expected that the more efficient pre-heat trains will employ a number of stream splits. Thus, a preheat train should benefit more than most plants from the availability of multi-stream units.

An alternative design based on multi-stream exchangers was developed for which the designs presented in Figures 3 and 4 had previously been proposed.

The design uses three compact units based on the surfaces reported by Kays & London(2). (Whilst, these surfaces may not prove suitable for this application, they have the advantage of being well characterised. The exercise was aimed at identifying potential benefit, not real design!).

One unit was positioned prior to the desalter, one between the desalter and the flash column, and one after the flash column. It would have been possible to design a single unit to handle the full duty. However, the design would have been very complex.

The unit prior to the desalter contacted the crude oil with the two upper column recycles, the HGO and LGO column recycles, and the three product streams (kerosene, HGO and LGO). It had seven sections, providing 6 idependent feed and draw-off locations. The unit would have been 8.36m high and 0.3m wide. It would have used 120 plates.

The second unit consisted of just one section. It contacted the crude oil with the upper, the HGO and the LGO column recycles and the kerosene stream. Its overall height was 5.35m. It was 0.3m wide and again used 120 plates.

The third unit again consisted of seven sections. It contacted the crude oil with the kerosene product, and four of the column recycle streams (upper, HGO, LGO and lower recycles). It has an overall height of 10.67m and was again 0.3m wide using 120 plates.

No attempt was made to optimise the design.

Given the geometries involved, the best orientation for the three units would appear to be vertical. The plant would therefore appear to consist of three tall slim exchangers, a desalter, a flash column and the main fractionation column. As such, the layout becomes very compact.

The combined volume of the three units is around 8 cu.m. Using conventional exchangers the network would have needed 36 units (some consisting of a number of separate shells) containing a total surface area of 10 000 sq.m. in order to achieve the same energy efficiency.

The savings from the use of multi-stream units in terms of expected exchanger capital, materials usage, piping costs, other layout associated costs and energy consumption are potentially large.

Application in Energy Saving Revamps

The energy saving revamp of crude trains has also been considered. A recently reported[3] retrofit technique based on 'path analysis' was used.

The network prior to retrofit is shown in Figure 7. The revised design is shown in Figure 8.

The retrofit involved the installation of around 3700 sq.m. of additional surface. The bulk of this (85%) was used in the two multi-stream units. The remainder was used in conventional two stream units required to compensate for tightened temperature driving forces.

The multi-stream units used in the revamp are much simpler than those used in the new plant design. The first unit was a four stream unit with the crude contacted with the fractionation column overheads, the top and upper column recycles.

The second unit was a three stream unit. The crude was contacted with the upper column recycle and the CGO stream.

The energy saving was 8.1MW out of a previous consumption of 33.4 MW (a saving of 25%). Since, cost data for multi-stream units does not exist, project payback cannot be determined.

Future Needs

The above studies have shown that multi-stream exchangers could have a very strong role to play in both grassroots design and plant retrofits.

The savings in new designs could be large.

Three and four stream units have a role to play in retrofits.

The design of multi-stream exchangers requires the careful balancing of heat loads and pressure drops. A procedure for plate-fin type units has recently been developed[4]. It is not clear whether or not such a procedure exists for plate-and-frame units.

Given that plate-and-frame units already have an established role in the process industries and readily lend themselves to three and four stream duties, a design procedure should be developed as a matter of priority.

These design procedures need to be supplemented by operability analyses in order to set to rest concerns regarding control and flexibility.

Acknowledgements

The above studies were conducted during part of a research studies funded by NOVEM, the Energy Efficiency Office and the following companies: AEA Technology, BP, British Gas, Exxon and Shell. All of this support is gratefully acknowledged.

The work was overseen by the Compact Heat Exchanger Study Group (chaired by Mr. J.Robertson) and the Netherlands Advanced Heat Exchanger Study Group (chaired by Ir. A.R.Braun). The contributions of these groups is again gratefully acknowledged.

References

1. Yee T.F., Grossmann I.E. & Kravanja Z. 'Simultaneous optimisation models for heat integration: 1. Area and energy targeting and modelling of multi-stream exchangers', Comp. & Chem.Engng. 1990,14(10),1151-1164

2. Kays W.M. & London A.L. Compact Heat Exchangers, McGraw-Hill, New York, 1984

3. van Reisen J., Grievink J., Polley G.T. & Verheijen P. 'The placement of two stream and multi-stream exchangers in existing networks through path analysis', Paper presented at ESCAPE 5, Slovenia, June 1995

4. Picon Nunez M. 'Development of Compact and Multi-stream Exchanger Design methodologies for Use in Integrated Plants' Ph.D. Thesis, U.M.I.S.T., 1995

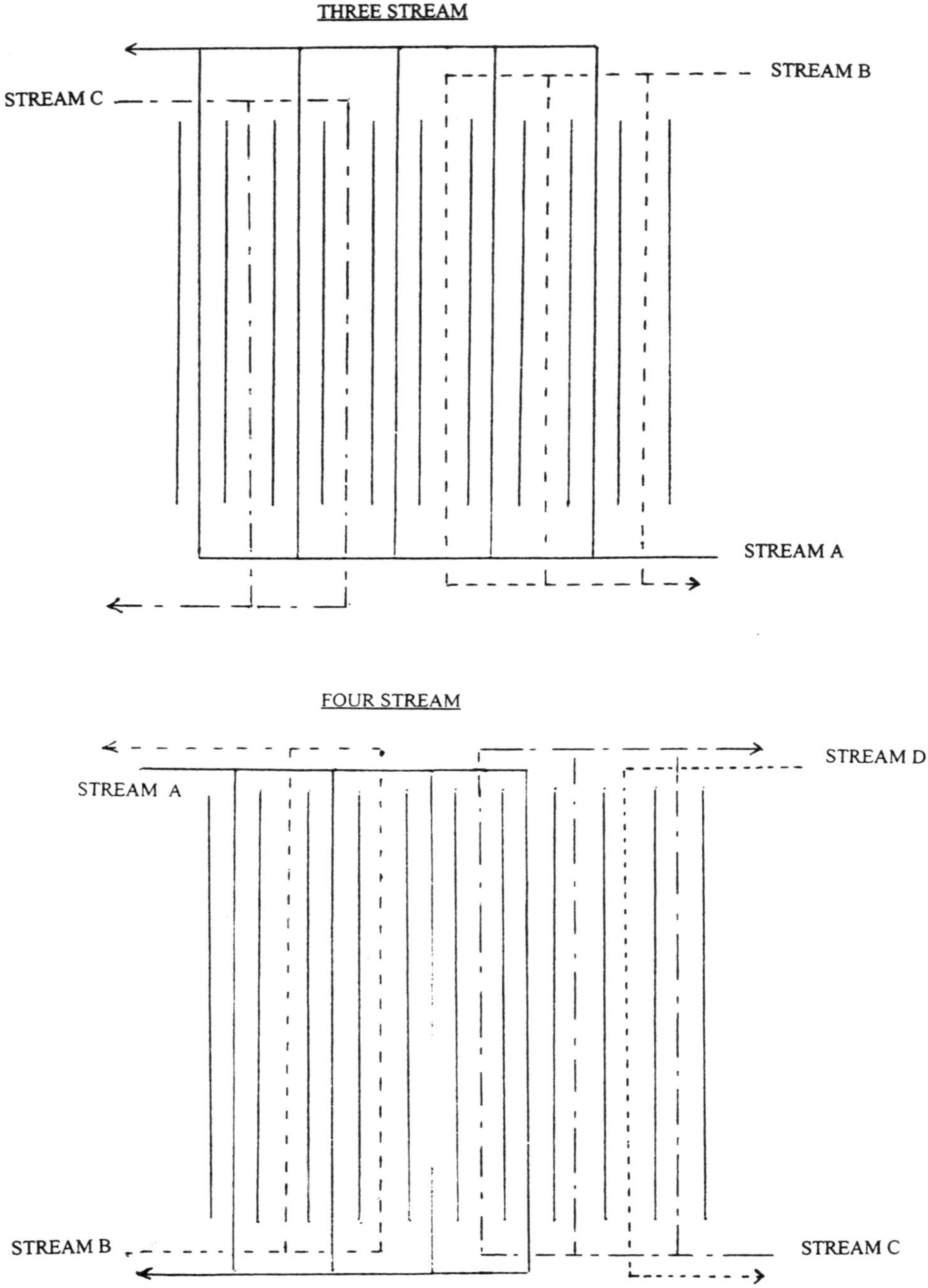

FIGURE 1. MULTI-STREAM PLATE-AND-FRAME UNITS

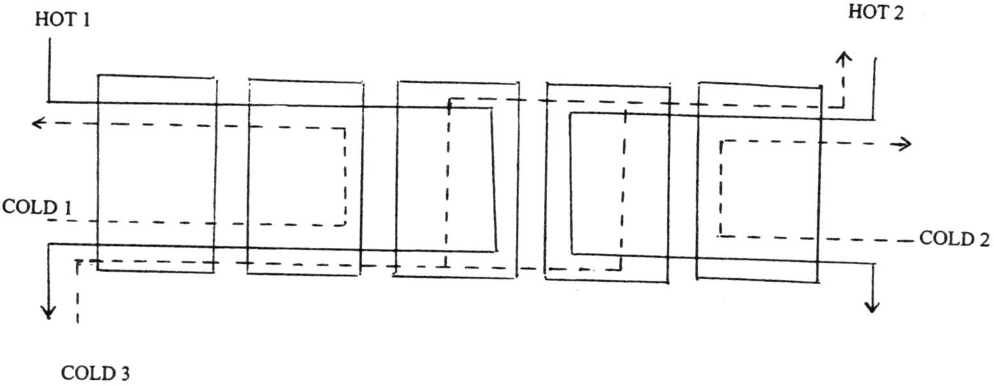

FIGURE 2. FIVE STREAM UNIT POSSIBLE IF SPECIAL PORTS INSTALLED

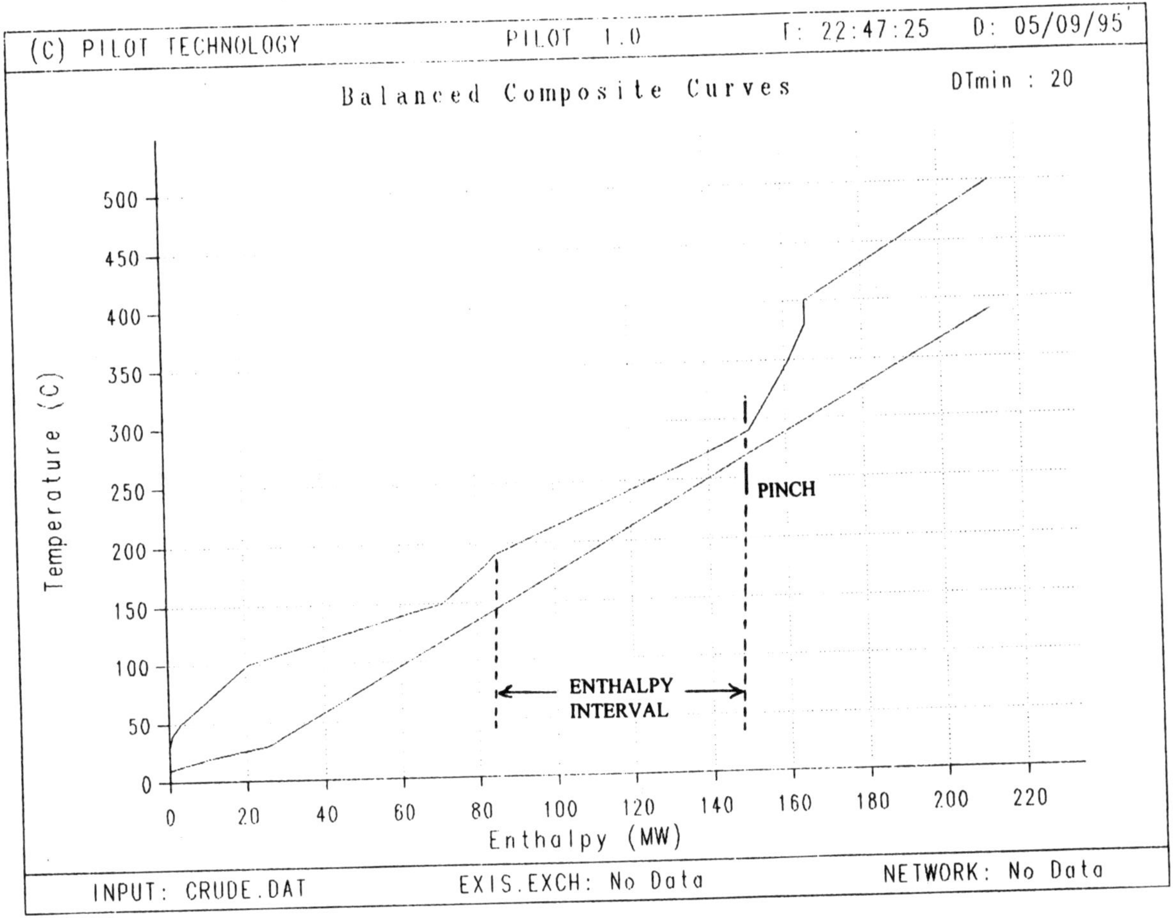

FIGURE 3. **COMPOSITE CURVES FOR CRUDE OIL PRE-HEAT TRAIN**

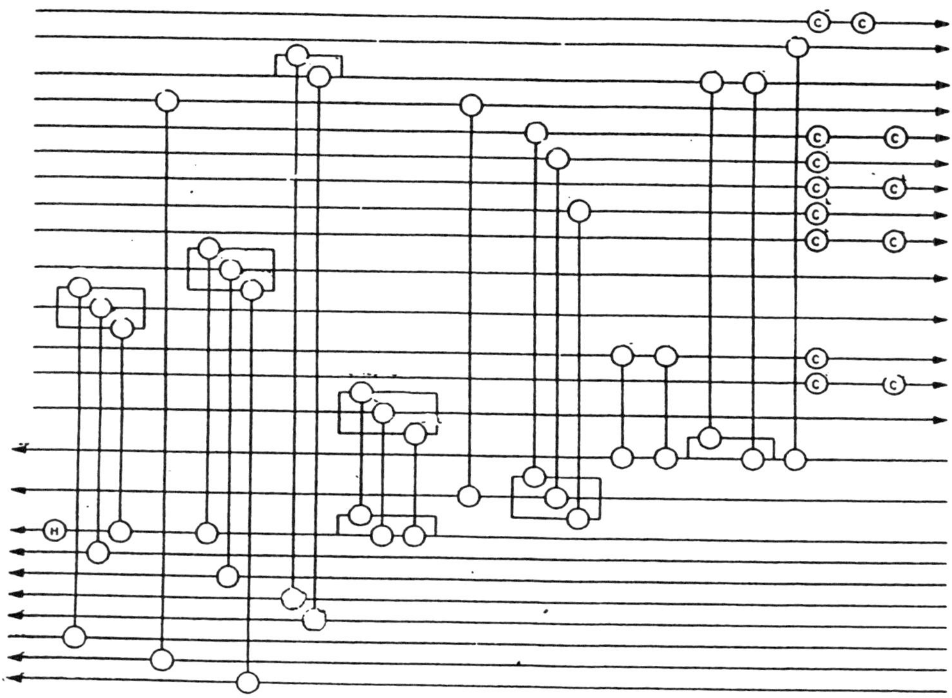

FIGURE 4. CRUDE PRE-HEAT TRAIN PRIOR TO PINCH TECHNOLOGY

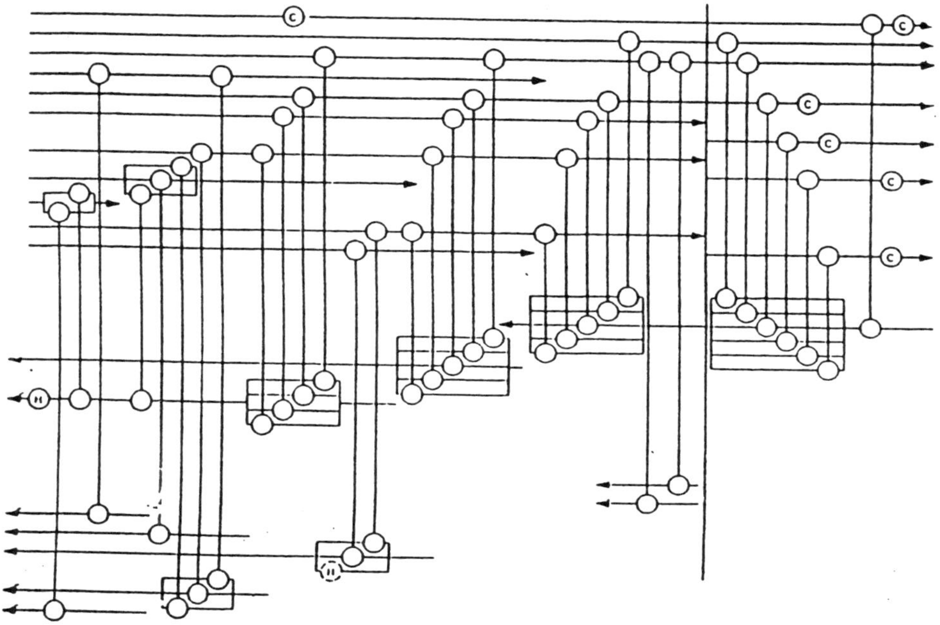

FIGURE 5 THE OPTIMUM NETWORK CONFIGURATION FOR THE BASE CASE (AFTER PINCH TECHNOLOGY)

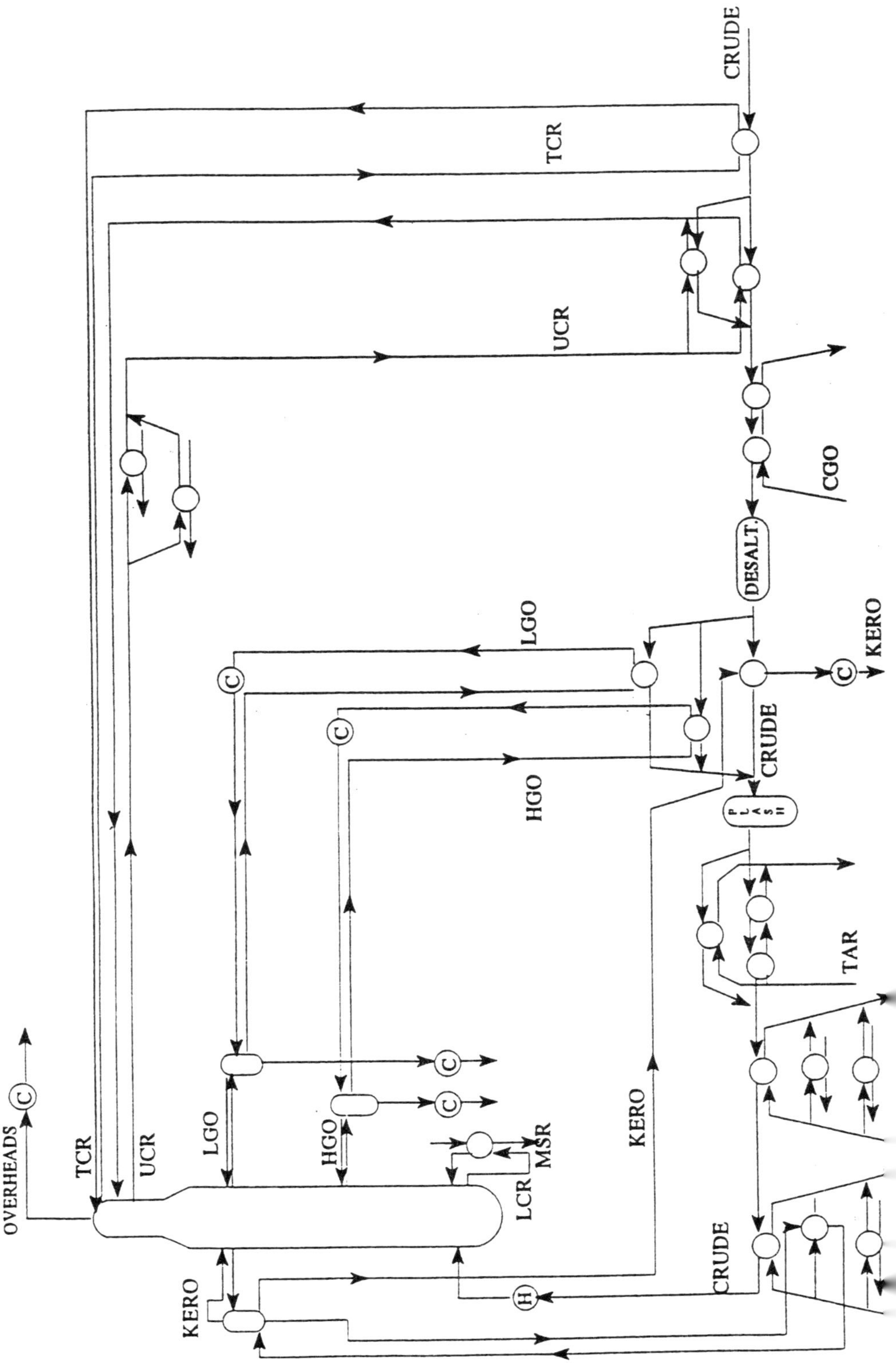
CRUDE
TCR
UCR
CGO
DESALT.
LGO
KERO
C
CRUDE
HGO
FLASH
TAR
OVERHEADS
TCR
UCR
LGO
HGO
MSR
LCR
KERO
CRUDE
KERO
H

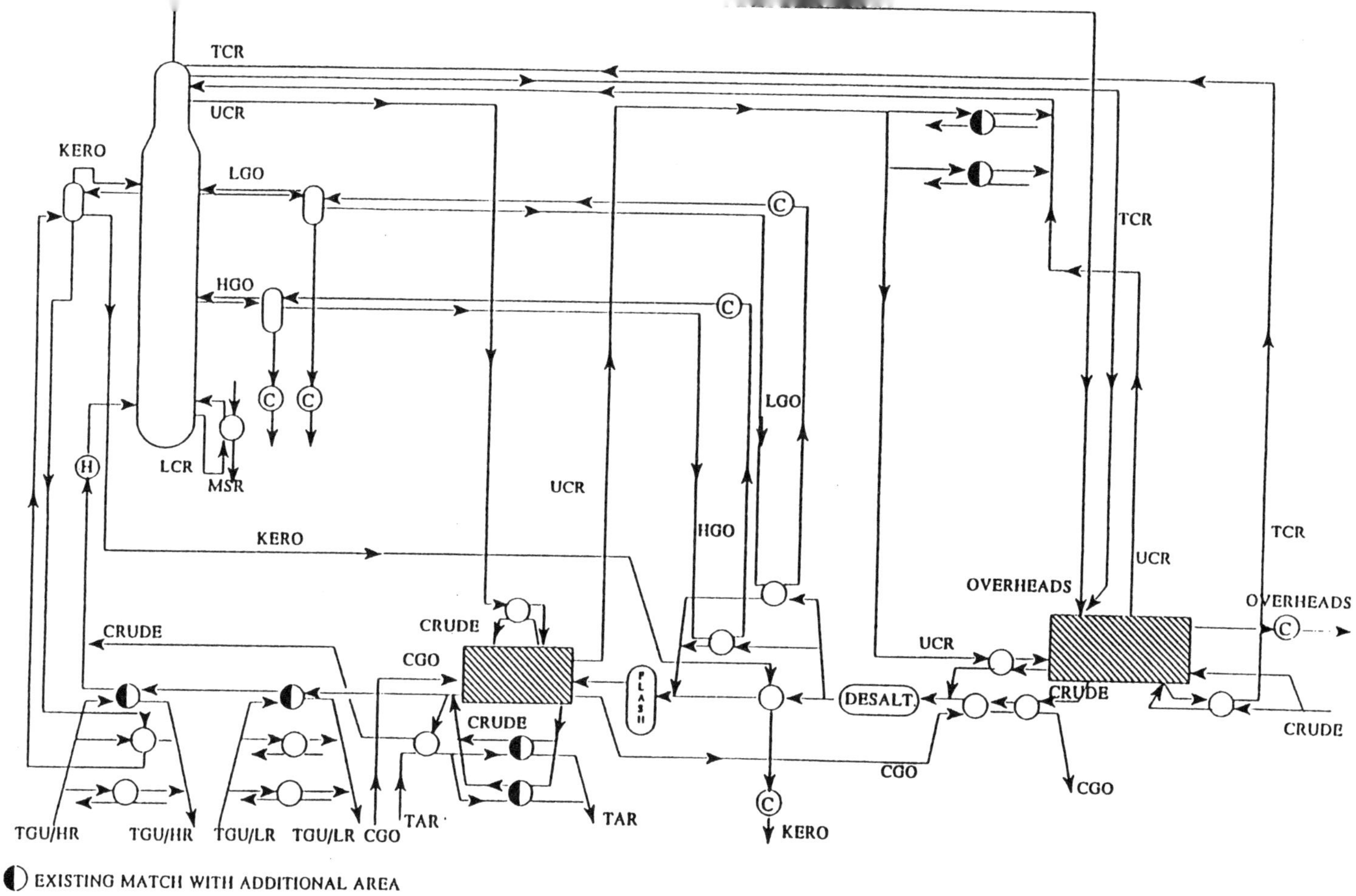

FIGURE 7. POSSIBLE RETROFIT

HIGH INTENSITY MIXING

ASSESSING ROTOR/STATOR MIXERS FOR RAPID CHEMICAL REACTIONS USING OVERALL POWER CHARACTERISTICS

T. G. Sparks[1],
D. E. Brown[2],
A. J. Green[1].

Summary

Rotor/stator mixers are used extensively in the process industries, but much less research work has been reported on them than on stirred tanks or motionless mixers.

This paper considers the overall power characteristics of a rotor/stator mixer so that local energy dissipation rates (and hence mixing rates) might be estimated. These are then assessed in process intensification terms.

---

[1]BHR Group Limited, Cranfield, Bedford, MK43 0AJ, UK.

[2]Biotechnology Centre, Cranfield University, Bedford, MK43 0AL, UK.

NOTATION

SYMBOL	MEANING	UNITS
K	constant (used in disk friction equation)	-
L_0	initial scale of segregation	m
N	shaft speed	s^{-1}
P_{SHAFT}	shaft power	W
P_{PUMP}	pumping power	W
P_{MECH}	mechanical losses	W
P_{DISK}	disk friction losses	W
P_{LEAK}	leakage (recirculation) losses	W
$P_{HYDRAULIC}$	hydraulic losses	W
Q	flow rate	$m^3 s^{-1}$
t_E	engulfment time	s
t_{MESO}	meso mixing time	s
Δp	differential pressure	Pa
η	pumping efficiency	-
ε	turbulent energy dissipation rate	$W kg^{-1}$
Λ	shaft torque	N m
ν	kinematic viscosity	$m^2 s^{-1}$

1 BACKGROUND

This section gives a general background to rotor/stator mixers, discusses previous work and defines the power balance that will be used to characterise the rotor/stator mixer.

1.1 Rotor/stator mixers

A rotor/stator mixer contains a high speed rotor spinning close to a motionless stator. Fluid passes through the region where rotor and stator interact and experiences highly pulsating flow and shear. In-line rotor/stator mixers resemble centrifugal pumps and may contribute significantly to pumping the fluid. This is shown schematically in Figure 1 and Figure 2 shows an industrial example.

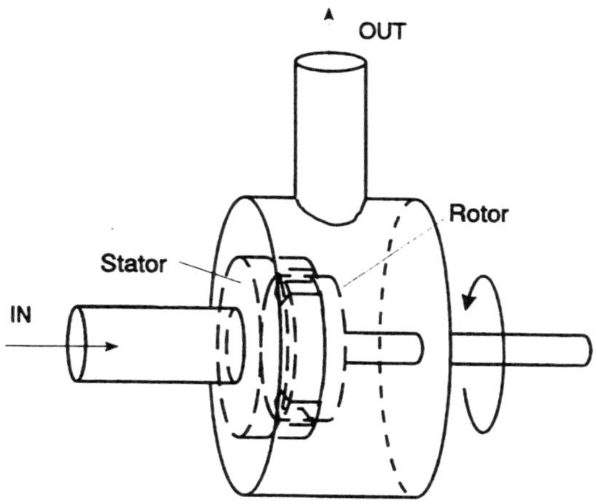

Figure 1 A rotor/stator mixer

Figure 2 An industrial example of a rotor/stator mixer

1.2 Process intensification

Process intensification (PI) is a design philosophy where the fluid dynamics of a process are matched to its chemical and physical requirements. For fast (particularly fast, competitive) chemical reactions, this implies short residence time reactors with intense mixing, and continuous or semi-batch operation. In-line rotor/stator mixers appear to offer such a combination of features.

1.3 Turbulent energy dissipation rate and mixing rates

Mixing rates can affect the outcome of competitive chemical reactions, where the intrinsic rate of the reaction is comparable to the rate of mixing (Ref. 1). Several mixing timescales can be defined, any one of which might be the rate limiting step.

Micro mixing - the stage of mixing that brings about final, molecular, homogeneity - depends upon the rate of dissipation of turbulent energy. For a low viscosity liquid the micro mixing rate is given by the rate of engulfment within small turbulent eddies (Ref. 2):

$$t_E \approx 17\left[\frac{\nu}{\varepsilon}\right]^{1/2} \quad \textbf{(1)}$$

Within larger turbulent eddies, 'meso mixing' (Ref. 3), reduces the scale of segregation of reactants (although not bringing about the molecular homogeneity needed for reaction - this requires micro mixing). A characteristic timescale (Ref. 4) is

$$t_{MESO} \sim \left[\frac{L_0^2}{\varepsilon}\right]^{1/3} \quad \textbf{(2)}$$

where L_0 is a characteristic length for the initial scale of segregation.

To define the mixing rate for a reactor, perhaps in order to model the performance, the equations above show that the turbulent energy dissipation rate per unit mass, ε, is a crucial parameter.

1.4 Previous work on rotor/stator mixers

Bourne and Garcias-Rosas (Ref. 5) and Bourne and Studer (Ref. 6), used the product distribution of competitive diazo coupling reactions conducted in a rotor/stator mixer to determine local turbulent energy dissipation rates. They report values for ε of up to 700 W kg^{-1} (Bourne and Garcias-Rosas) and 1200 W kg^{-1} (Bourne and Studer). These correspond to a micro mixing time (in water at 25°C), t_E, of $\sim 2\times10^{-3}$ to 1.5×10^{-3} s, sufficiently rapid to compare with the intrinsic speed of fast competitive chemical reactions, and therefore affect their outcome.

The technique used gave local information, the diazo coupling reaction scheme can be used to infer average ε in the reaction zone. They report that removal of the stator has no effect on the outcome of these reactions - implying that the reaction is over before the shear gap (the thin region between the rotor and stator) or the stator is reached, or that the stator has no effect on the ε.

This paper attempts to infer energy dissipation rates by estimating ε in terms of measurements of the overall power balance.

1.5 Power balance for a centrifugal pump

The following power balance was adopted (Ref. 7) for a centrifugal pump and can be applied to a rotor/stator mixer:

$$P_{SHAFT} = P_{PUMP} + P_{MECH} + P_{DISK} + P_{LEAK} + P_{HYDRAULIC} \tag{3}$$

Figure 3 shows the power balance for a double suction centrifugal pump (from Ref. 8). The components of this balance are discussed below.

P_{SHAFT} - This is the shaft power input to the mixer.

P_{PUMP} - The pumping power output, this would be needed to assess whether an additional pump is needed.

P_{MECH} - The mechanical losses, in Figure 3, are the losses due to friction in the seals and bearings. Some definitions of mechanical loss include disk friction, which is not included here because the seal and bearing losses could be measured separately from the disk friction losses. Mechanical losses (and the disk friction loss) do not contribute to mixing.

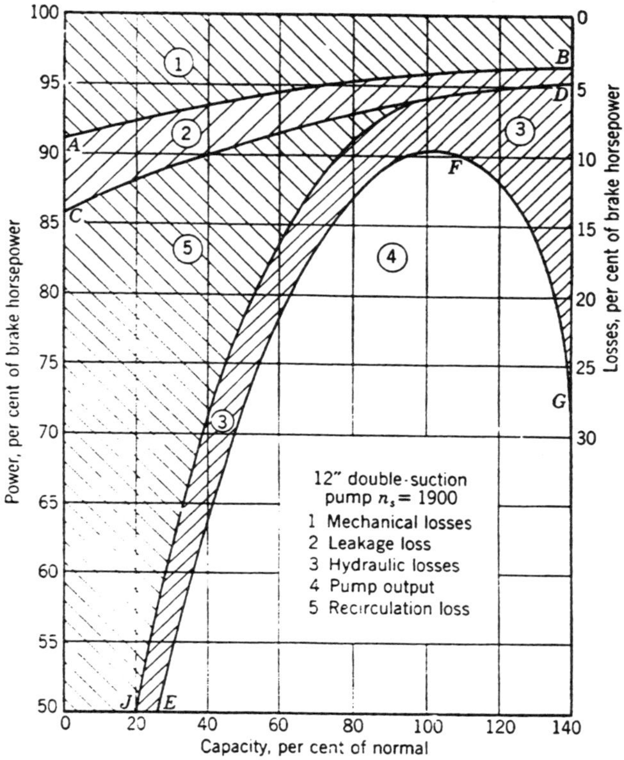

Figure 3 Power balance for a double suction centrifugal pump (Ref. 8)

P_{DISK} - Disk friction losses are caused by frictional drag on the solid surfaces of the rotor, and are analogous to the skin friction losses in a motionless mixer. Disk friction losses are significant in pumps and they are accounted for in this work.

P_{LEAK} - The leakage loss can mean the loss due to leakage through seals or, as it is in this work, the loss due to fluid not passing through in the required direction. An example of this type of leakage (or recirculation) loss is fluid passing back from the high pressure (downstream) side of the impeller to the low pressure (upstream) side. This is a macro mixing mechanism (reducing coarse scale concentration fluctuations) and will influence the residence time distribution.

$P_{HYDRAULIC}$ - Hydraulic losses (turbulent losses) are the losses that contribute to micro mixing.

2 EXPERIMENTAL DETAILS

2.1 Rig/ methods

A rotor/ stator mixer, based upon a Silverson machine, was placed in a flow loop - see

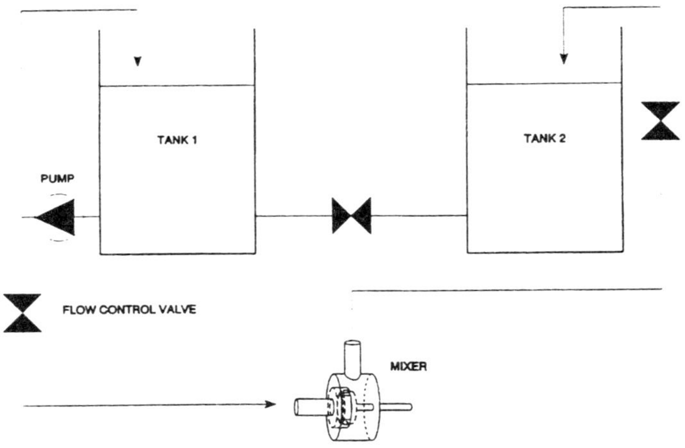

Figure 4 Test loop (Ref. 9)

Figure 4. Shaft torque, shaft speed and differential pressure were measured. The machine contained a mixer unit designed by the author to model a typical toothed rotor/stator type, see Figure 5. The rotor diameter was 121.45 mm and the inner diameter of the stator teeth was 123.8 mm, both rotor and stator had 18 teeth and a tooth open area of 50%.

Figure 5 Toothed rotor/stator

The following paragraphs show how components of the power balance, see Equation 1, were measured or estimated.

Shaft power, P_{SHAFT}, was determined by measuring the shaft torque and the rotational speed and calculated using:

$$P_{SHAFT} = 2\pi N\Lambda \tag{4}$$

The pumping power output of the mixer, P_{PUMP}, was found by measuring the flow rate and differential pressure:

$$P_{PUMP} = Q\Delta p \tag{5}$$

Mechanical losses were obtained by measuring the shaft torque and shaft speed with the rotor removed, P_{MECH} was then calculated using Equation 5.

Disk friction losses, P_{DISK}, were determined using an empirical expression given in pumping text books (Ref. 10):

$$P_{DISK} = KN^3D^5 \tag{6}$$

It was not possible to measure hydraulic and leakage losses ($P_{HYDRAULIC}$ and P_{LEAK}) directly with the methods outlined here. It was assumed that the remainder of the power balance (not accounted for above)is the hydraulic and leakage losses combined:

$$P_{HYDRAULIC} + P_{LEAK} = P_{SHAFT} - (P_{PUMP} + P_{DISK} + P_{MECH}) \tag{7}$$

3 RESULTS/ DISCUSSION

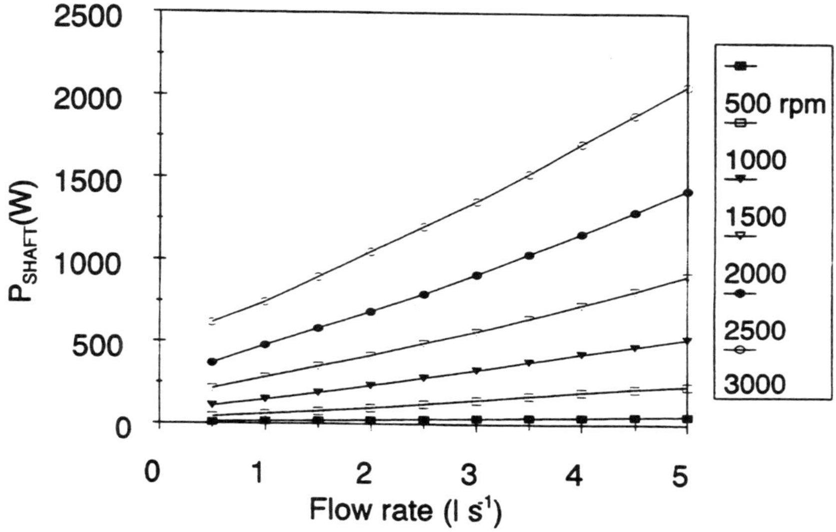

Figure 6 Shaft power against flow rate for a range of shaft speeds

Figures 6 and 7 show shaft power and pumping power data, and Figure 8 shows the overall energy balance for the mixer at 3000 rpm.

Figure 6 shows that shaft power rises approximately with flow rate and that, at maximum flow rate and shaft speed the shaft power is approximately 2 kW. The proportion of this power going to turbulence and mixing will be considered later. Each curve in Figure 7, the pumping power curves, shows a maximum; in fact at low shaft speeds and high flow rates there is a nett pressure drop across the mixer. This is because the differential pressure across the mixer shows the combined effects of pressure rise due to the rotor and the 'static' losses that occur in the system.

3.1 Power balance

The power balance, Figure 8, summarises the power dissipation and transformation in the mixer at 3000 rpm. It shows that the mechanical and disk friction losses account for

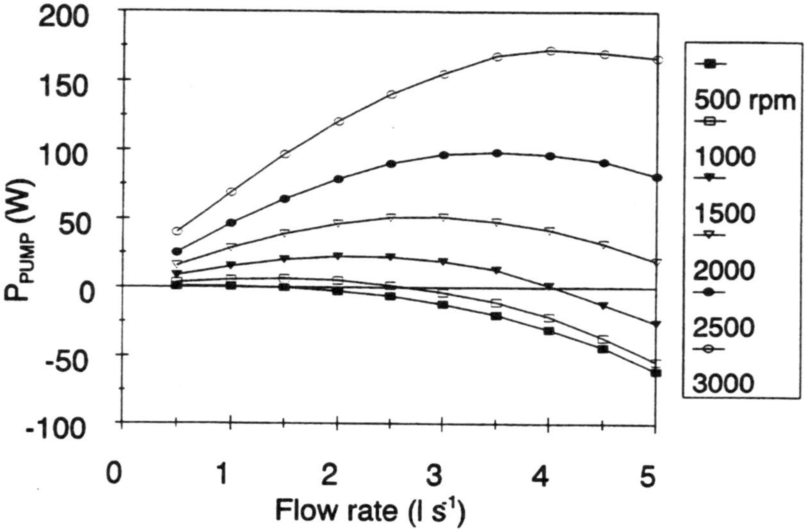

Figure 7 Pumping power against flow rate

about 25% of the shaft power input at low flow rates but at high flow rates, the overall shaft power increases. The mechanical and disk friction losses (which do not depend on flow rate) therefore account for a lower proportion of the total shaft power input. The remainder of the power balance is discussed in the following sections.

3.1.1 Pumping efficiency

The pumping efficiency of this unit is very much lower than that for a centrifugal pump (it was not designed to pump):

$$\eta = \frac{Q\Delta p}{P_{SHAFT}} \qquad (8)$$

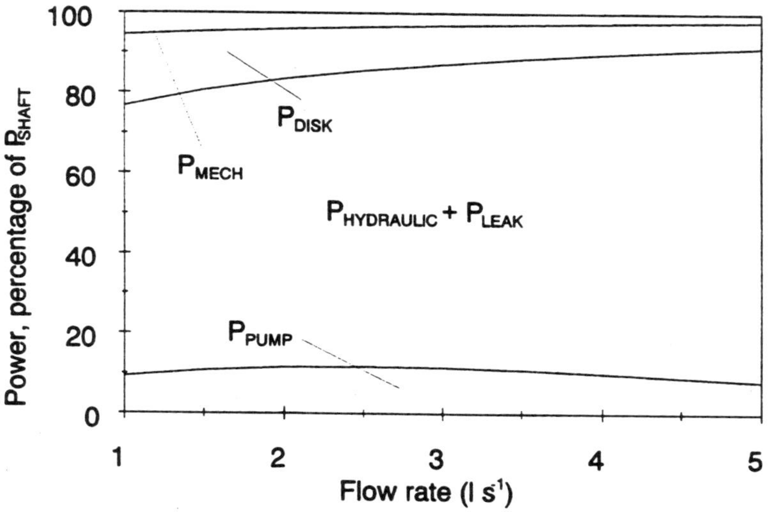

Figure 8 Power balance at 3000 rpm

With centrifugal pumps this value might be as high as 90%. In the power balance shown here it is ~10-15% at the best efficiency point.

There is a basic hydrodynamic reason for the best efficiency point (BEP) in a pump, which occurs when the tip speed of the rotor matches the speed of the flow in the volute. In the rotor/stator mixer, where the rotor and the volute are effectively separated by the stator, there is not a simple set of conditions that accompany the BEP.

3.1.2 $P_{HYDRAULIC}$ and P_{LEAK}

Most of the shaft power input (~70%) is not accounted for by the other components of the power balance, it is therefore concluded that this power is dissipated through hydraulic and leakage losses combined.

Using these techniques it is not possible to measure ε directly. It is assumed that approximately 1/3 of the hydraulic and leakage losses combined is dissipated to turbulence and that this occurs in the immediate vicinity of the rotor/stator teeth.

At 3.0 l s^{-1} and 3000 rpm, the shaft power is approximately 1.2 kW (Figure 6). If 1/3 of this (400 W) is dissipated through turbulent energy dissipation within the rotor/stator teeth (in this case fluid mass ~0.07 kg) this gives ε~5500 W kg^{-1}.

This value, an estimate averaged over the extent of the rotor/stator teeth and including the shear gap, is somewhat higher than that obtained using the chemical reaction technique used by Bourne and co-workers (in smaller mixers). This could be explained by their result that the removal of the stator had no effect - it is probable that energy dissipation rates in the shear gap and stator are higher than those in the rotor.

Further, the energy balance shows that the proportion of power going to hydraulic and leakage losses is relatively insensitive to flow rate. The power balance for the pump showed that the hydraulic losses (component 3 in Figure 3) increase rapidly as the best pumping efficiency point is passed. In fact the BEP is not as well defined as for the pump. The drop in efficiency above the BEP arises due to increasing 'static' losses in the mixer (in fact, at high flow rates there could be a nett pressure drop across the mixer).

3.2 Implications for mixing performance

Turbulent energy dissipation rates of 5500 W kg^{-1} correspond to a micro mixing time (in water) of 7×10^{-4} s and a meso mixing time (acting upon an initial scale of segregation of 1 mm and calculated using Equation 2) of ~6×10^{-4} s.

These indicate extremely rapid mixing and, provided it is known how best to exploit these mixing rates (feed tube location etc.), then rotor/stator mixers will create suitable conditions for the intensification of processes.

3.3 Conclusions

i Rotor/stator mixers are inefficient pumps (10 - 15%).

ii Approximately 70% of the shaft power input to a rotor/stator mixer is dissipated through hydraulic and leakage losses.

iii This proportion is relatively insensitive to small changes in flow rate (it appears to depend more upon shaft speed), so the mixer can be used in a loop where small changes in flow rate may occur.

iv At 3000 rpm ε, the turbulent energy dissipation rate per unit mass is estimated to be approximately 5500 W kg^{-1}. This corresponds to a micro mixing timescale of approximately 7×10^{-4} s in water at room temperature.

v The rotor/stator mixer would appear to be a suitable device for chemical reactions with intrinsic rates ~10^{-3} s.

4 REFERENCES

1. **Baldyga, J. & Bourne, J. R.** "A Fluid Mechanical approach to turbulent mixing and chemical reaction." Part II Micro mixing in the light of turbulence theory. Chem. Eng. Commun 28 (1984). pp 243-258.

2. **Baldyga, J. & Bourne, J. R.** "Simplification of Micro mixing Calculations I. Derivation and Application of New Model." Chem. Eng. J 42, 83-92 (1989). pp 83-92.

3. **Baldyga, J., Bourne, J.R., Dubuis, B., Etchells, A. W. Gholap, R. V. and Zimmerman, B.** "Jet reactor scale-up for mixing controlled reactions." Trans IChemE, Vol 73, Part A, July 1995. pp 497-502.

4. **Corrsin, S.** "The isotropic turbulent mixer: part II. Arbitrary Schmidt number." A.I.Ch.E.J vol 10 (1964). p 870.

5. **Bourne, J. R. & Garcias-Rosas, J.** "Rotor Stator Mixers for Rapid Micro mixing." Chem. Eng. Res. Des, Vol 64. (1986). pp 11-17.

6.	**Bourne, J. R. & Studer, M.**	"Fast Reactions in Rotor/stator Mixers of Different Size." Chem. Eng. Process 31, (1992). pp 285-296.
7.	**Anderson, H. H.**	"Centrifugal pumps." 3rd edition. Trade and technical press. ISBN 85461 076 6. (1980).
8.	**Stepanoff, A. J.**	"Centrifugal and axial flow pumps, theory design and application." 2nd edition. John Wiley and Sons ISBN 0-471-82137-3. (1957).
9.	**Sparks, T. G.**	"Power balance and residence time distribution tests on a rotor/stator mixer." HILINE interim report number HLP05, BHR Group Limited. December 1994.
10.	**Karassik, I. J. Krutzsch, W. C. Fraser, W. H. and Messina, J. P. eds.**	"Pump Handbook." 2nd edition. McGraw -Hill, New York. ISBN 0-07-033302-5. (1986).

EXPERIMENTAL STUDY OF THE TURBULENT FLOW INSIDE A MOTIONLESS IN LINE MIXER

A. Mokrani*, C. Castelain, H. Peerhossaini

Thermofluids and Complex Flows Research Group
Laboratoire de Thermocinétique URA CNRS 869
Université de Nantes
ISITEM La Chantrerie CP 3023
F44087 NANTES Cedex 03 - France
**e-mail: asen@isitem.univ-nantes.fr*

Abstract

The aim of this experimental work is to study heat-transfer mechanisms in turbulent flow through a motionless in-line mixer. The fluid is intended to undergo an exothermic chemical reaction, and mixing enhances both the chemical process and the heat transfer. The first step in the work takes a fluid-dynamics approach. In this paper, we present the results of flow visualizations performed in this geometry that reveal the longitudinal coherent structures predicted by previous computations.

I INTRODUCTION

In-line reactors/heat exchangers are widely used in the chemical, food-processing and pharmaceutical industries. For reactors in which the flow is turbulent, it is important to understand the controlling length scales of the turbulent spectrum limiting the reaction in progress. The objective of our experimental study is to model the reacting flow through the heat exchanger passage (called "test section" henceforth). The role of the test section is to generate longitudinal vortical structures in the fluid flow. The fluid is intended to undergo an exothermic chemical reaction, and efficient mixing enhances both the chemical process and the heat-transfer mechanisms. In addition, the product composition of the chemical reaction depends strongly on the fluid temperature, which is governed by heat-transfer phenomena. In the first step of this study we take a fluid-dynamics approach: the test section is made of perspex and has flat outer walls to permit flow visualizations and LDA measurements.

In the next section, we present a detailed description of the experimental setup. The basic flow and some preliminary results of flow visualization experiments are described in section 3, and section 4 presents some conclusions.

II DESCRIPTION OF THE EXPERIMENTAL SET UP

The perspex test section (see figure 1) consists of a straight tube of inner diameter 20 mm along which seven tab arrays are fixed. Each of the seven arrays consists of four trapezoidal tabs positioned at 90° to each other and fixed on the tube wall. The tabs are turbulence promoters and generate longitudinal vortical structures. The test section is 180 mm long and the distance between two successive tab arrays is 20 mm (one tube diameter).

The test section is connected to a preconditioner and a postconditioner, which are 300 mm straight transparent tubes of circular inner cross-section and 20 mm inner diameter (the same as that of the test section). The preconditioner is used to provide a fully developed flow at the inlet of the test section, and the postconditioner to observe the effect of the test section on mixing quality. Care was taken that the connections between the test section and both preconditioner and postconditioner did not introduce any geometrical perturbation in the flow.

A schematic diagram of the experimental setup is shown in figure 2. The working fluid is water. In order to reduce flow fluctuations, the test section is fed by an open loop in which an overhead reservoir is filled to a constant level by city water. Water flows from this reservoir by gravity through a flexible tube connected to the preconditioner. The water temperature is measured by a platinum probe to determine its physical properties (viscosity, density) and to calculate the Reynolds number. The flow rate is controlled by a valve and measured by a bank of three flowmeters. The error due to these flowmeters is less than 1.6%. After flowing through the test section and postconditioner, the water is evacuated to the drain.

For flow visualizations, a tracer dye is injected parallel to the flow direction through two injectors of 0.25 mm inner diameter, one injector located 10 mm upstream of the first tab array and the other 2 mm downstream of the third tab array. Along the tube diameter, the radial positions of both injectors can be adjusted within ± 0.01 mm by two micrometer traversing mechanisms. However, for all visualizations presented here we have used only the injector located upstream of the first tab array, the radial position of which was maintained 0.5 mm from the tube wall. The flow cross-section is illuminated by an argon ion laser. The laser light crosses a silica bar to generate a laser sheet that is then thinned by a planocylindrical lens. When illuminated by a laser light at 514 ηm wavelength, the fluorescein tracer absorbs an amount of energy that is re-emitted at a higher wavelength. This phenomenon, called laser-induced fluorescence (LIF), allows visualization of the flow cross-section only; other light reflections are filtered using an appropriate optical filter.

The cross-sections of longitudinal vortices are then visualized by a JVC 3CCD high-resolution camera. Both the camera and the laser sheet generator device are fixed to a rigid frame so that no relative movement between them is possible. The frame is mounted on a traversing mechanism. An electronic displacement ruler allows determination of the position of both camera and laser sheet within 0.02 mm. The abscissa of the laser sheet along the test section can be read on a digital liquid crystal display. The rig lets one follow the longitudinal evolution of the cross-section of streamwise vortices very closely. Pictures from the camera can be recorded on a Sony Umatic video cassette recorder (VCR) and can then be played back, processed by microcomputer and digital image-processing software, printed, etc. All the techniques used for flow visualization are described in detail in [ref. 1] and [ref. 2].

III EXPERIMENTAL RESULTS

Experiments have been run for Reynolds numbers between 1500 and 3500. We present here the visualizations for Reynolds number 1500. Figure 3 shows a pair of longitudinal vortices visualized 2 mm downstream of one of the first four array tabs. As shown in the figure, the height of the vortices is of the same order as the height of the tab. At this Reynolds number, the flow in the smooth circular preconditioner is laminar and the tracer dye flows along a thin straight line before reaching the first tab array. Downstream of the first tab array, video visualization shows that the pair of vortices remains stable until the second tab array is reached.

Y. Leroux and J. Bodé [ref. 3] have computed the velocity field in the geometry studied here. Their results show that when the fluid flows along a tab, the pressure on the top of the tab is greater than that beneath it. This generates a pair of counter-rotating streamwise vortices downstream of the each tab (see figure 4) that transport fluid particles from regions close to the tube wall towards the tube centerline, thus promoting radial fluid mixing.

Along the second tab array, the dye is pushed up towards the tube centerline and begins to become unsteady. Figure 5, a photograph of the flow cross-section visualized 2 mm downstream of the second tab array, shows that a new pair of vortices is generated by the second tab array. The vortices generated by the first tab array are still discernible, but they have been pushed up towards the tube core by the new vortices. At this position, video visualizations show that the structures are no longer steady but begin to undergo apparently random oscillations. The dye is not yet engulfed by the vortices generated by the other tabs of the same array.

Figure 6 shows a photograph of the flow cross-section visualized 2 mm downstream of the third tab array. At this position, the vortices generated by the third tab array do not appear very clearly, since the tracer dye begins to spread into the whole tube cross-section. Video visualization shows that the tracer dye is trapped in vortices generated by other tabs of the same array. Organized structures can still be seen, although they appear intermittently.

A visualization of the flow cross-section 2 mm downstream of the fourth tab array is shown in figure 7. At this position the tracer dye is no longer trapped in the vortices created by a single tab, but occupies almost the whole cross-section of the tube. No organized structure can be continuously observed any longer. Video visualization shows that the structures oscillate at higher frequencies, in a random way. At this position, the flow appears turbulent but homogeneous mixing is not yet achieved.

Figures 8 and 9 are photographs of the flow cross-section visualized 2 mm downstream of the fifth and sixth tab arrays, respectively. Here, again, organized structure can no longer be seen: the tracer dye occupies the whole cross-section and mixing is further enhanced.

IV CONCLUSION AND FUTURE WORK

The preliminary flow visualizations presented here were carried out at Reynolds number Re = 1500. Photographs of the flow cross-section were taken at different positions along the test section. In a straight smooth tube, the tracer dye would follow a straight streak-line, since the flow is laminar. The photographs give a qualitative idea of the mixing enhancement due to the presence of the tabs. Coherent structures have been visualized as well and confirm the numerical predictions of Y. Leroux and J. Bodé [ref.3]. However, since these structures oscillate, video visualization is better suited to observing them.

The next step in this study will be to measure the flow velocity field; in addition, turbulent quantities such as $\overline{u'v'}$, $\overline{u'w'}$, $\overline{v'w'}$ and turbulent energy dissipation ε will be evaluated. At the same time, a quantitative characterization of mixing will be made by applying the mean gradient of relative concentration (MGRC) technique [ref. 4] to the visualization images, and the results will be related to the turbulent energy dissipation.

ACKNOWLEDGMENTS
This work has been supported by European Union under Joule II program. The project was coordinated by British Hydrodynamics Research Group Ltd.

REFERENCES

1 Y. Le Guer, H. Peerhossaini, *"Order breaking in Dean flow"*, Phys. Fluids A3 (5), 1991.

2 H. Peerhossaini, C. Castelain, Y. Le Guer, *"Heat exchanger design based on chaotic advection"*, Exp. Therm. and Fluid Sci. 1993, 7:333-344.

3 Y. Leroux and J. Bodé, *"CFD study on the HEV mixer, Joule II Program"*. Personal communication, 1995.

4 P. Guillemet, F. Danes, H. Peerhossaini, *"A new method to characterize a mixing state"*, Proceedings of the Eurotherm 39 seminar on "Heat transfer enhancement by Lagrangian chaos and turbulence",p 57-63. Nantes (France) 1994.

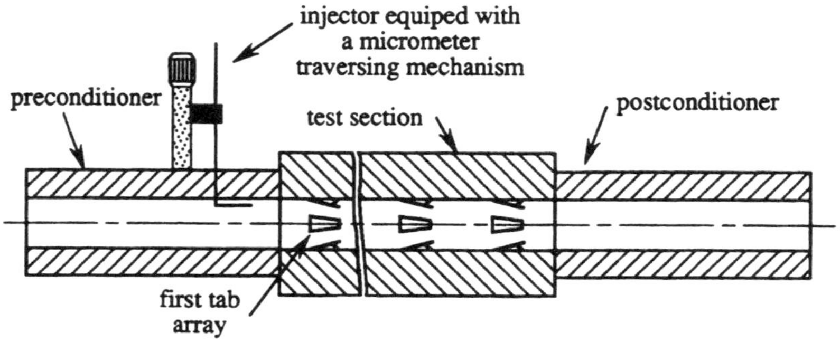

Figure 1 : Test section, preconditioner and postconditioner

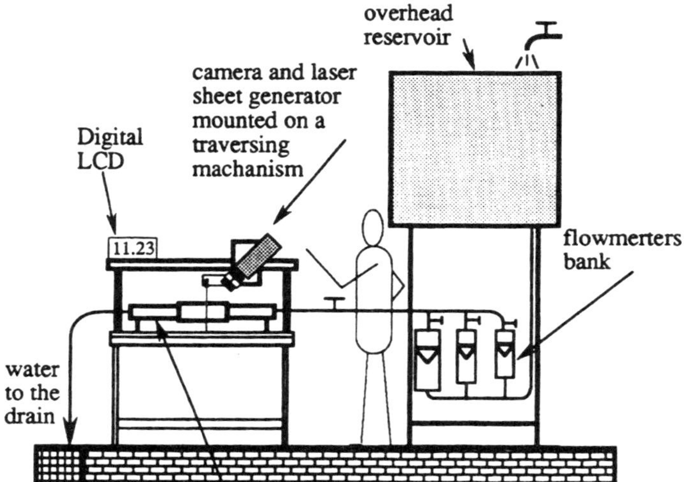

Figure 2 : Schematic diagram of the experimental setup

Figure 3 : Visualization of the flow cross-section
2 mm downstream of the first tab array

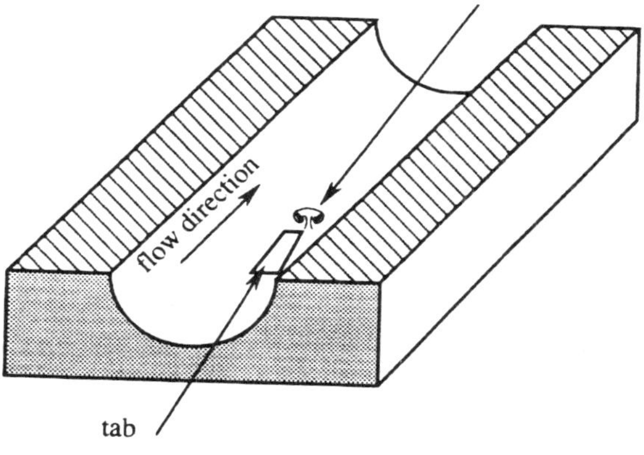

Figure 4 : Vortex generation by a tab

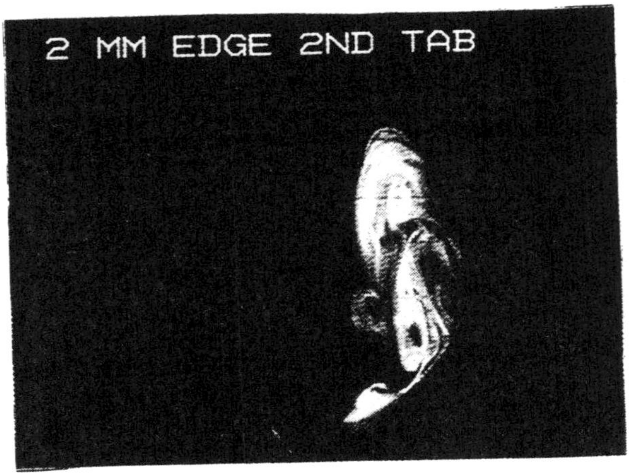

Figure 5 : Visualization of the flow cross-section
2 mm downstream of the second tab array

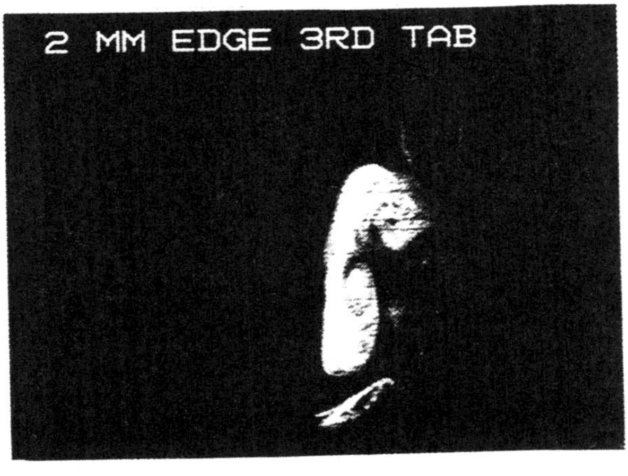

Figure 6 : Visualization of the flow cross-section
2 mm downstream of the third tab array

Figure 7 : Visualization of the flow cross-section
2 mm downstream of the fourth tab array

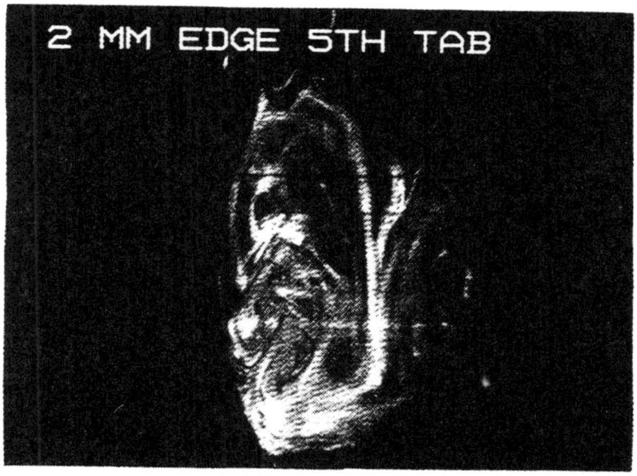

Figure 8 : Visualization of the flow cross-section
2 mm downstream of the fifth tab array

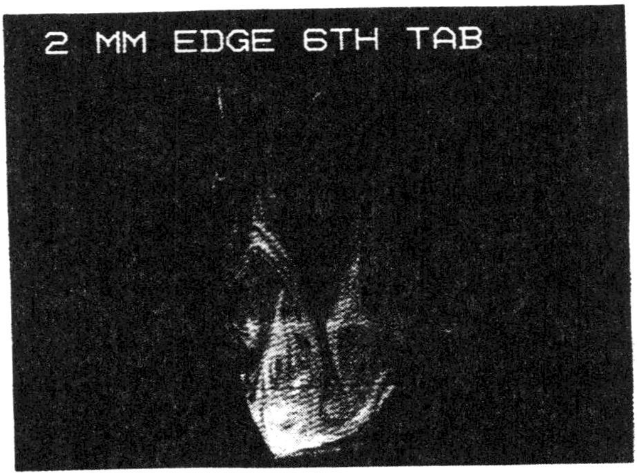

Figure 9 : Visualization of the flow cross-section
2 mm downstream of the sixth tab array

A NUMERICAL STUDY OF EXOTHERMIC REACTIONS IN A WALL COOLED STATIC MIXER

Dipl.-Ing. Götz Lauschke
Hoechst AG
Verfahrenstechnik, G 811
D-65926 Frankfurt
Tel.: +49 69 305 1 58 25
Fax: +49 69 30 90 14
email: LSK@MSMFZJ.HOECHST.HOECHST-AG.D400.DE

ABSTRACT - A numerical study on reaction intensification in a HEV motionless in-line mixer with wall cooling is presented. The simplified test system consists of an aqueous solution. CFD computations employing a k,ε-model with wall functions describe qualitatively well the inhomogeneous turbulent flow with recirculation zones and counter-rotating vortices. Heat transfer simulations for a hot liquid and a cooled wall indicate temperature patterns resulting straightforward from inhomogeneous flow. An exothermic generic chemical reaction with fast kinetics is considered. The reacting species are fed separately into the reactor. Assuming only macromixing a reaction zone is formed mainly in the fluid core.

1 INTRODUCTION

A crucial topic of process intensification in the chemical industry is designing reactors which combine both excellent mixing and heat transfer properties. Static mixers have proved principally their potential to fulfil these requirements. However, for specific process and reaction conditions the selection of the appropriate equipment represents a nontrivial task. This is due to the complex interaction of flow field, (exothermic) chemical reactions, and heat transport. CFD simulations are particularly suitable to predict at least qualitatively these phenomena.

The high efficiency vortex (HEV) mixer provides inhomogeneous turbulence by vortex generation with symmetrical tabs fixed at the wall (Fig. 1). This arrangement results in a low pressure drop because the fluid core in the centre is not deviated by flow obstacles. *Fasano, 1991* has reported HEV mixing performance data for a wide range of Reynolds numbers including thermal mixing efficiency and pressure drop. In the present study simultaneous mixing, chemical reaction and heat transport are systematically analysed by CFD computations. In order to discriminate the various phenomena combined, calculations of the isolated processes have been performed. First, the flow field is analysed, then focus is on heat transfer from the fluid to the wall without chemical reaction. The temperature field gives also important information on macromixing. Finally, results are presented for a simple exothermic reaction in a wall cooled HEV. The present study is intended as a first step on the way to compare different mixing models and several degrees of detail in reaction modelling.

2 MODELLING

The mixer section considered here consists of a tube (d = 20 mm) with 5 tab arrays fixed in intervals of 20 mm. Each tab array contains four trapezoidal tabs regularly arranged on the perimeter. The angle between tabs and wall is 30°, tab length is 7 mm, length of trapezium top is 4 mm , and length of trapezium bottom is 7 mm. The computational grid starts 10 mm in front of the first tab array and ends 10 mm behind the last. An injection tube (d = 7.3 mm) for one feed stream is located in the centre of the mixer, the other feed stream is fed through the remaining cross section area. In this way a feed ratio of 7.5 is set, assuming equal feed velocities. For Reynolds number 15000 the velocities are 0.6675 m/s. If no asymmetric swirl is introduced through the reactor inlet, symmetry can be exploited to generate the three dimensional computational mesh as shown in Fig. 2 with cyclic boundary conditions on the radial areas. Coarse grid calculations on a one tab interval have pointed out spatial domains where rigorous grid refinement is required. In Fig. 2 the resulting fine grid is presented.

For turbulence modelling a high Re k,ε-model with standard wall functions (*Launder, Spalding, 1974*) is applied. The fluid consists initially of water at a temperature of 298 K. The physical properties are kept constant. For the heat transfer simulations the feed temperature is set on 368 K whereas constant wall temperature is at 298 K. The heat transfer coefficient amounts $\alpha = 500\ W/m^2K$.

A simple generic reaction scheme is applied

$$A + B \rightarrow C + D$$

in order to discriminate the single reaction step clearly from the interacting phenomena. The first order Arrhenius type kinetics in A and B are associated with a pseudo rate constant $A = 6\ 10^{10}\ s^{-1}$ and an activation energy E_A = 46 kJ/mol. If the reaction time scale is of the same order of magnitude as the micromixing time scale or even faster, micromixing models may describe the dependence of conversion from the degree of mixing (*Bourne, 1984, Villermaux, 1992*). The governing micromixing time scale for fluids with Sc >> 1 (i.e. liquids) is the Batchelor time scale for the smallest concentration fluctuation eddies. It demotes

$$\tau_B = \left(\frac{D}{\varepsilon}\right)^{1/3} \qquad (1)$$

with molecular diffusivity D and energy dissipation ε. For inhomogeneous flows energy dissipation becomes $\varepsilon = \varepsilon(x,y,z)$. Therefore τ_B also exhibits a non-negligible spatial pattern. In multicomponent reacting flows species may have different diffusivities. In this case τ_B can show additional inhomogeneities. For global kinetics and isothermal reactions the reaction time scale can be expressed in terms of the lowest reaction rate constant. However, in industrial reactors with heat release and cooling, the reaction time scale as well has to be considered as a spatial parameter. The HEV in conjunction with the reaction discussed represents such a system where local data is required. Consequently, the first modelling step reported here consists in accounting for macroscopic transport and reaction. The continuing step will be the implementation and testing of micromixing models. This procedure enables a rigorous comparison between several modelling approaches. From the industrial viewpoint it is particularly interesting to determine the necessary degree of detail in reaction modelling.

3 SIMULATION RESULTS

For flow and heat transfer simulations feed consists only of pure water. Assuming temporally constant vortex structures at Re = 15000, a steady state flow field as depicted in Fig. 3 is computed. The top picture represents the flow field on the plane x = 0 in Cartesian coordinates.

The simulations indicate axial recirculation zones developed directly under the tabs. These axial recirculation zones have properties of dead zones with large residence times even at high Re numbers. At the end of the reactor section the velocity field is fully developed in all spatial directions. The phenomenon governing macroscopic mixing consists in a pair of counter-rotating streamwise vortices depicted in Fig. 3 bottom. From one tab array to the next the vortices induce swirl in the flow field so that a uniform pattern is achieved behind five tab arrays. This evolution is documented at three axial positions (ends of tabs 1, 3 and 5) in Fig. 3 bottom. At the last tab array the rotation velocity has increased and the vortices have grown to their maximal size. These results are in qualitatively good agreement with the experiments reported by *Mokrani, Castelain and Peerhossaini, 1995* and the visualisations communicated by *Hearn, 1994*.

The flow core from centreline to the top of the tabs is spatially accelerated up to double feed velocity. Steep velocity gradients are detected between core and recirculation zone behind the tabs. These steep gradients downstream from the top of the tabs generate regions of high local energy dissipation (Fig. 4). However, beside these small regions and wall boundary layers, energy dissipation is rather low in the HEV and segregation effects due to fast chemical competitive reactions may occur. The pictures at the bottom of Fig. 4 show further that energy dissipation attains its maximal value at the trapezoidal edge of the tabs. Note that the dissipation regions in the planes z = constant represent the rotating vortex pattern. This result may encourage to redesign the tabs if higher dissipation and more intensive micromixing is required.

The computations in Fig. 5 reveal both macromixing properties and heat transfer in the HEV. Feed in both inlets is at homogeneous temperature of 368 K. Heat transfer with constant wall temperature takes place at the entire wall. Nevertheless, no significant wall cooling is detected upstream from the first tab array. However, downstream in the first recirculation zone elevated residence times lead to lower temperatures. Macromixing of the counter-rotating vortices produces exchange with hot core fluid which results in a mixing temperature in front of the second tab array. This axial pattern continues creating distinct thermal mixing zones upstream from each tab. More complex patterns are computed at the z-planes of tab array 1, 3 and 5. From these pictures the main macromixing directions can be derived. Behind the first tab array the small vortices produce an elliptical thermal mixing field. The pair of vortices in Fig. 3 (bottom, left) is larger than the mixing field in Fig. 5 (bottom, left) because they lay still in the axial recirculation zone. Only upstream from this axial recirculation zone an important exchange with bulk fluid is computed. As flow develops radial mixing from tab to centreline is preferred to mixing on the perimeter. Along the tab underside small quantities of cool fluid are transported by the axial vortex into the macromixing region. At the last tab array there exist still regions close to the centre and far from the tabs near the wall showing almost inlet temperature. These are critical regions from the reaction engineering viewpoint.

If a fast chemical reaction is involved, a reaction zone is formed reminding of a turbulent diffusion flame. The reactants A and B are separately fed into the reactor. A passes through the injection tube, B enters through the remaining inlet cross section area. The reaction starts spontaneously from the inlet at the interface of the feed streams. Macromixing is sufficient to achieve high conversion in the HEV section.

4 CONCLUSION AND PERSPECTIVES

The HEV computations presented here were performed for Re = 15000. This is in the operation range recommended by the manufacturer. Flow is in turbulent regime under these conditions, so a standard k,ε-model with wall functions has been employed. The simulated flow field is qualitatively in good agreement with experiments carried out by several researchers. Experimentally observed patterns have been predicted by the calculations. Heat transfer simulations show macromixing properties of the HEV and indicate distinct temperature pattern.

Future work will focus on the implementation of micromixing models in order to compare the performance of these models with macroscopic predictions. A further more detailed approach will also include the pdf approach for chemically reacting systems.

ACKNOWLEDGEMENTS

This work has been supported by European Union in the framework of Joule II program. The project is coordinated by British Hydrodynamics Research Group Ltd.

REFERENCES

J. R. Bourne ``*Micromixing revisited*'', Inst. Chem. Eng. Sympos. Ser. 87 (1984), ISCRE 8, p 797

J. B. Fasano ``*Kenics HEV Mixer Sets a New Standard for Turbulence Mixing Efficiency*'', Mixing XIII, Banff Alberta, Canada, 10 - 15 June 1991

S. Hearn ``*Visualisation notes of a dye blending experiment in a HEV*'', unpublished, British Hydrodynamics Research Group, 1995

B. E. Launder, D. B. Spalding ``*The numerical computation of turbulent flows*'', Comp. Meth. in Appl. Mech. & Eng. 3 (1974), p.269

A. Mokrani, C. Castelain and H. Peerhossaini ``*Experimental study of the turbulent flow inside a motionless in line mixer*'', 1st Int. Conf. on Proc. Intens. for the Chem. Ind., 6.-8.12.1995, Antwerp

J. Villermaux ``*Mixing in polymer reactors*'', Dechema Monographien 127, VCH Verlagsgesellschaft, Weinheim 1992

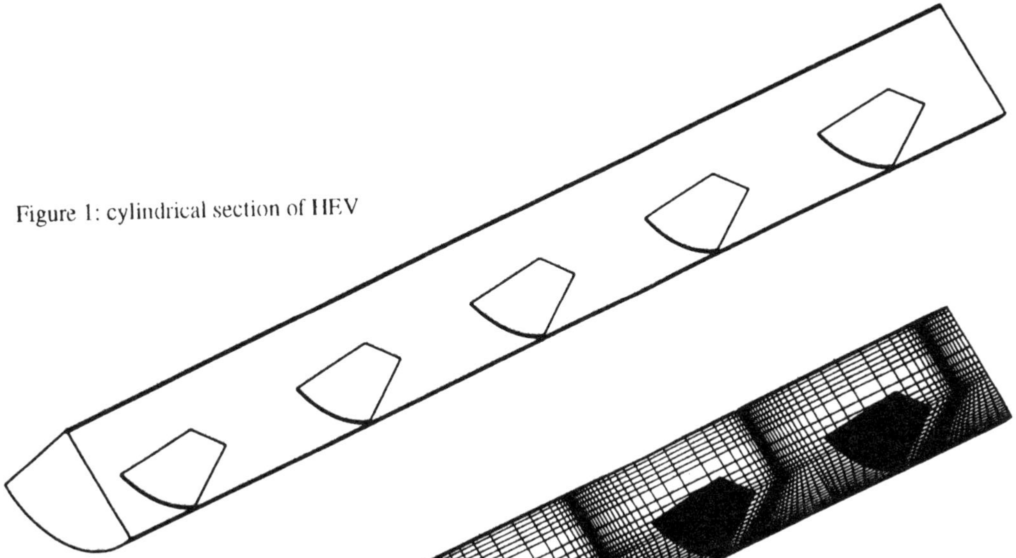

Figure 1: cylindrical section of HEV

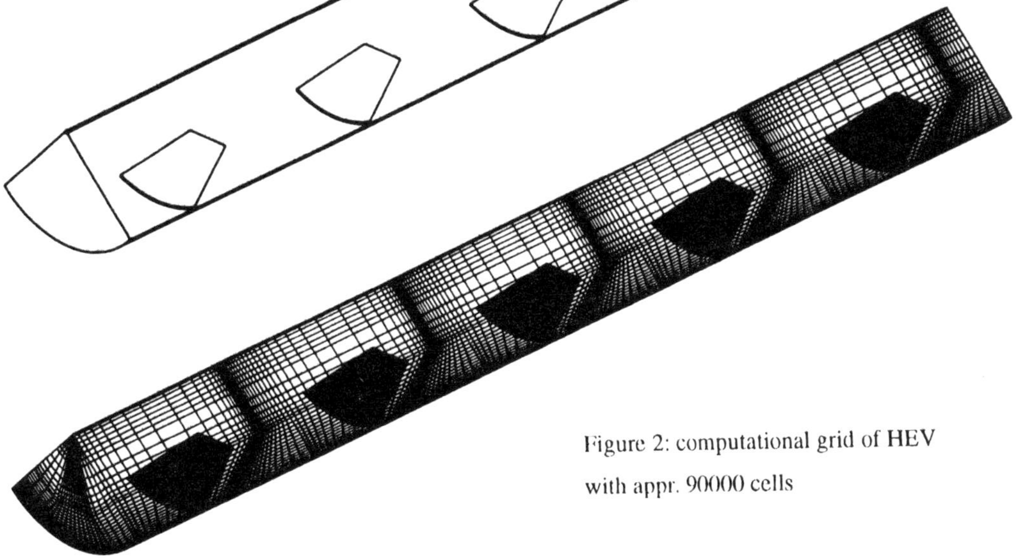

Figure 2: computational grid of HEV
with appr. 90000 cells

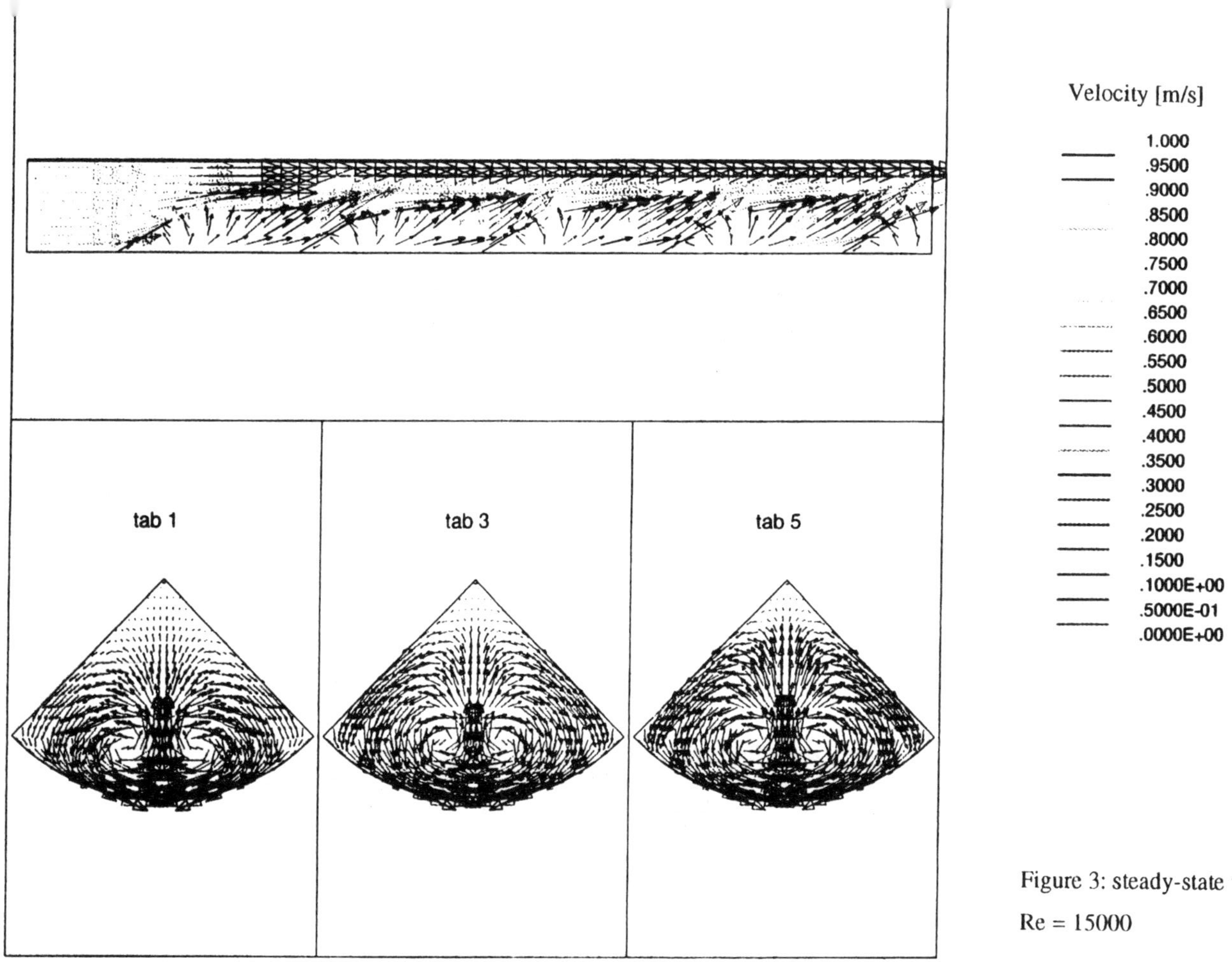

Figure 3: steady-state velocities for Re = 15000

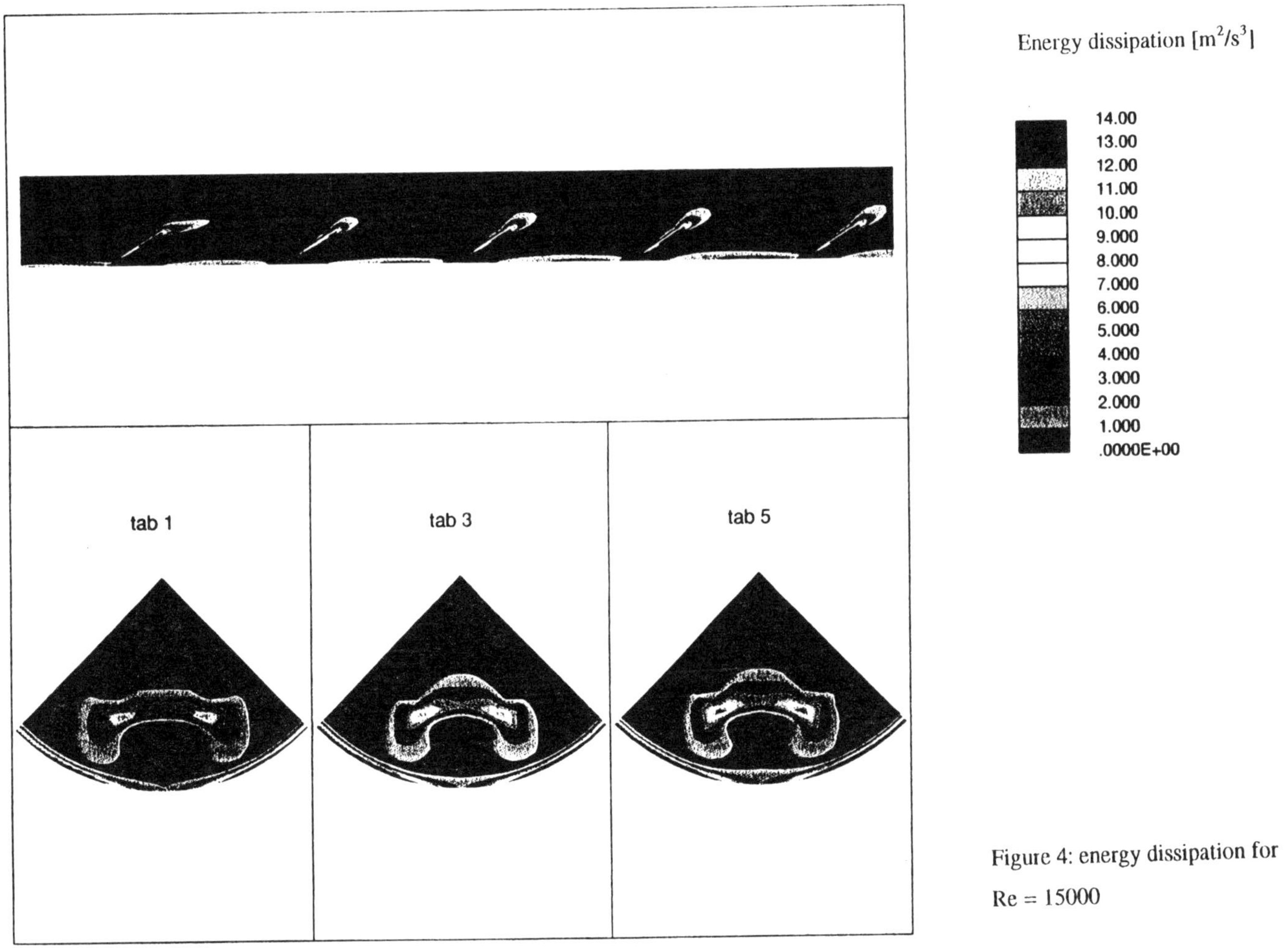

Figure 4: energy dissipation for Re = 15000

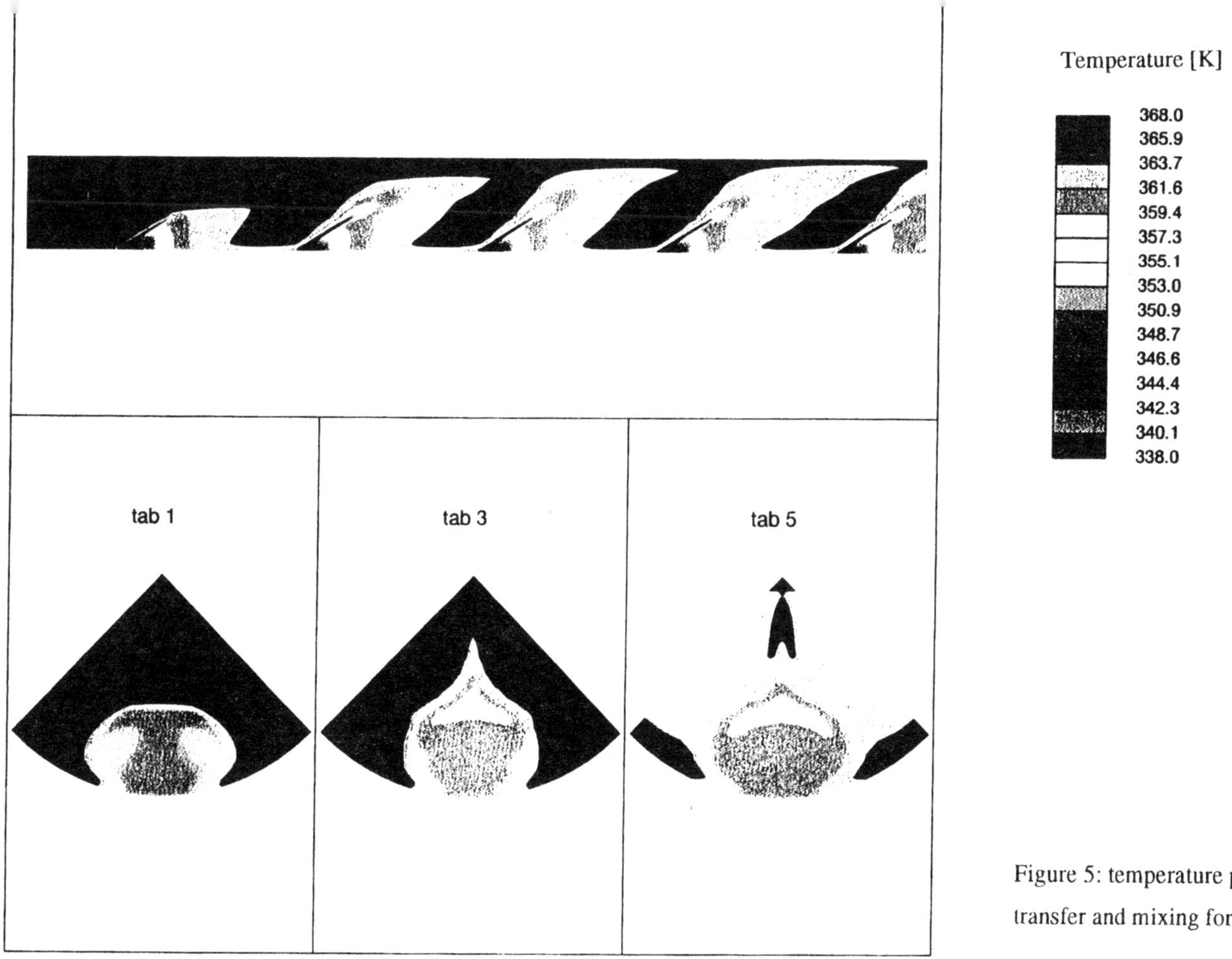

Figure 5: temperature pattern due to heat transfer and mixing for Re = 15000

SEPARATION

AND

MEMBRANES

Use of Cross-Flow Nanofiltration Membrane Technology in Process Intensification and Plant De-bottlenecking

K.Abhinava, ZENECA Ltd., UK

Abstract

Traditionally, the manufacture of fine chemicals has involved synthesis, product recovery and finishing (down-stream) technologies. This approach can lead to poor yields and less than optimum use of synthesis and down-stream assets in a number of product areas. This paper describes the use of cross-flow nanofiltration technology in overcoming the limitations of the traditional process for a specific range of products. This relatively new technology has allowed significant increases in the synthesis and downstream capacities, increased the yields and significantly reduced the environmental impact of the process. This paper particularly addresses the process development aspects of the new technology.

Cross-flow nanofiltration is applicable where separation of large molecules (> 1000 MW) is required from small molecules (< 200 MW). Separation is also possible if there are charge differences (i.e. divalent ions from monovalent ions etc.). As the capital and running costs of nanofiltration plants are high, this further restricts general applicability. However, in the fine chemicals industry, NF technology is finding increasing use.

1. Traditional Process

Figure 1 shows a schematic of the traditional process for the manufacture of a multi-product range of fine chemicals using batch processing. Following synthesis the product is precipitated (isolated) requiring careful control and extensive optimisation of the isolation process to obtain the product in a suitable form for further processing. Precipitation is carried out by addition of chemicals at suitable rates and concentrations. The product is then recovered by filtering the slurry using conventional filtration equipment. The filter cake is then fluidised and re-slurried before carrying out the downstream processing. The reason for re-slurrying the filter-cake is that it makes the product formulation and down-stream processing easier.

The traditional process suffers from the following disadvantages :-

a) Use of additional vessels for product isolation.

b) Expense of raw materials required for the isolation process.

c) Yield losses during the filtration process which also lead to significant environmental impact.

d) Use of additional vessels for re-slurrying the isolated product.

Figure 1 : Traditional Process

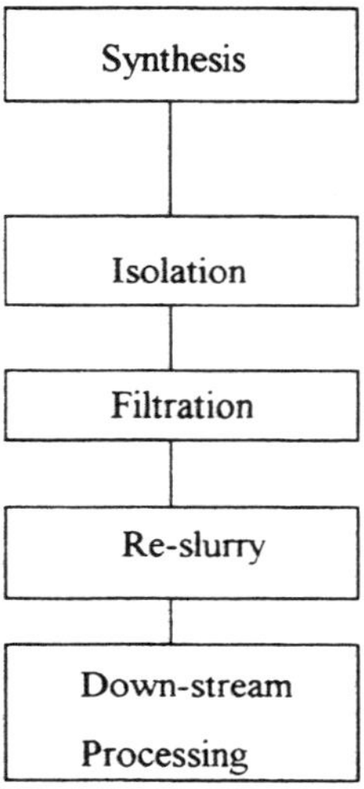

Figure 2 : Developed Process

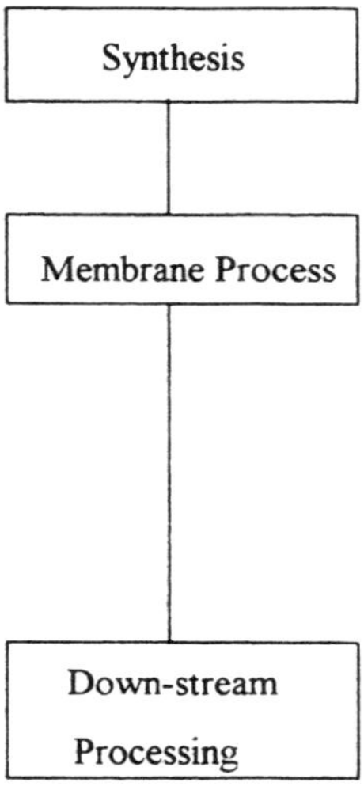

2. The Developed Process

The disadvantages of the traditional process have been overcome by the development of a new process involving the use of cross-flow *nanofiltration* technology.

Figure 2 shows the new process route. Following synthesis, the reaction mass is processed via a cross-flow nanofiltration (NF) unit. The NF membranes retain the product and allow the passage of low molecular weight species (organic and inorganic species as well as water). In simple terms, separation can be assumed to take place by the difference in the molecular weights of the product and other species and the pore size distribution characteristics of the NF membrane.

Following the purification and concentration by the membrane process the batch is then sent forward to the down-stream processing stages. The advantages of the new process include :-

a) A simpler process due to the replacement of three unit operations (product isolation, filtration, re-slurrying) with one.

b) Synthesis capacity improvement as the vessels previously used for the isolation process could be used for synthesis.

c)The isolation process was also the bottleneck step in the batch processing and therefore the elimination of this step also helped in lowering batch cycle times and thus improving the synthesis capacity.

d) As the bottleneck moved upstream to a synthesis stage, this opened up the possibility of telescoping the processes by combining two or more stages at the new bottleneck stage.

e) Savings on raw materials used in the isolation process.

f) Yield improvement due to an order of magnitude lower product losses via the membrane unit compared to the filtration process.

g) Significant environmental benefits due to the above.

h) Increased capacity of downstream processing due to less complexity and lower vessel volume requirements.

i) Achievement of consistent feed to the down-stream processing which resulted in quality improvements. This consistency is achieved by the fact that the purification and concentration levels can be accurately controlled during the membrane process.

3. Process Development

3.1 Process Selection

The traditional process was a well established and robust process for product isolation and purification. Fundamental changes to the process were limited as the product degrades at high temperatures and high and low pH. Use of traditional separation methods could not be employed due the combination of one or more of the following factors - technical, commercial, effect on product strength, quality and stability.

The above factors led to the identification of a membrane separation technique as a possible technique for development.

3.2 Membrane Selection

The separation required was that of high molecular weight (1,200 to 2,000) product species from low molecular weight (<500) organic and inorganic impurities. Some commercially available membranes were tried. Some experimental membranes were also developed in conjunction with the membrane manufacturers. A number of membranes were discarded because the product losses were too high and/or the rates were very low and/or the membranes were not robust.

Initially, membranes with a nominal molecular weight cut-off (MWCO) of 500 to 1200 were tried. However, because of the complex mechanisms of nanofiltration separation, membranes with MWCO in excess of 2,000 were also tried.

The selection of membranes was eventually narrowed down to two commercially available membranes both giving the required performance. The membrane eventually selected had a greater tolerance of working conditions (pH, temperature and pressure) than the other.

Despite improvements in the reporting of pore-size/ retention/ permeate flux characteristics by membrane manufacturers, membrane selection remains an area where trials are necessary to identify a suitable membrane for a particular application.

3.3 Lab. Scale Trials

Following the selection of the membranes, lab. scale trials were carried out on the selected range of products to determine the optimum operating parameters and the quantification of benefits of the new process versus the traditional process for each product.

3.3.1 Operating Parameters

The operating parameters which effect the cross-flow nanofiltration process are pressure (P), temperature (T) and cross-flow velocity (cfv).

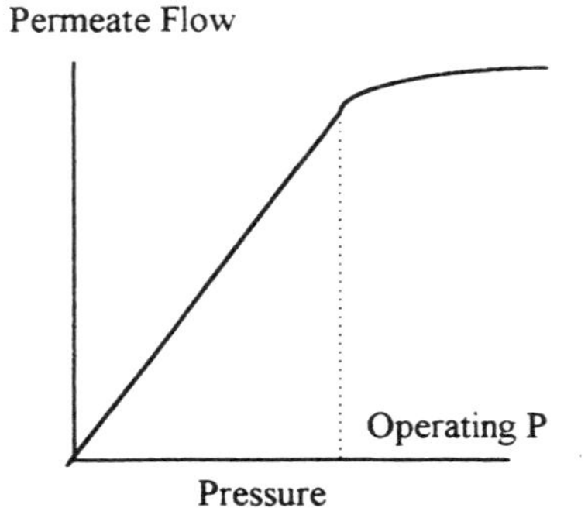

Fig 3.1 : Pressure versus Permeate Flow

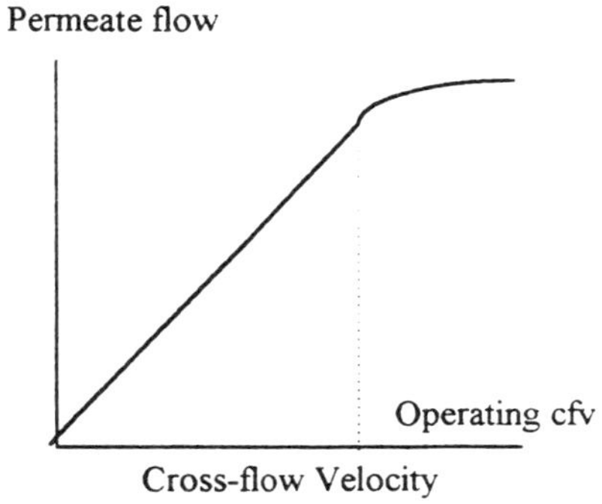

Fig 3.2 : CFV versus permeate flow

Figures 3.1 and 3.2 show the selection of the operating conditions for the membrane process. With increasing pressure the permeate flow rate increases proportionally until a pressure is reached beyond which any increase in pressure does not lead to any further increase in the permeate flow. This point was chosen as the operating pressure. Cross-flow velocity versus permeate flow shows a similar curve and cfv was chosen in a similar manner. The permeate flow rises proportionally with the temperature. The operating temperature selected was 40°C which was based on the known corrosive nature of the product at high (>50°C) temperatures.

3.3.2 Operating Regimes

For each product the optimum operating regimes were also determined at this stage. The membrane process is operated in batch recycle mode. There are two basic operations during the membrane process - *diafiltration* and *concentration*. Diafiltration is the operation by which the required level of purification is achieved. During diafiltration the permeate volume flow is replaced by solvent (water in this case) addition to the batch tank thus reducing the amount of impurities. Concentration step is a straight-forward volume reduction step. For each product, the optimum operating regime involved a combination of one or more of the diafiltration and concentration steps.

3.3.3 Quantification of Benefits

With the lab. scale trials, the advantages of the membrane process over the traditional process could be quantified from the data obtained for the product loss in the permeate and the concentrations and the purification achieved.

3.3.4 Product Trials / Cleaning Procedure

It was important to try all the products targeted for the membrane process as each had slightly different chemical synthesis route and it was not clear how robust the membrane process would be to varying types and amounts of any potential fouling species.

It was also important at this stage to develop an effective cleaning procedure for the membranes to reduce irreversible fouling and thus prolong membrane life. Proprietary chemical cleaning agents were tried and an effective chemical cleaner was found.

3.3.5 Membrane Configuration

At the lab. scale trial stage the selection of the membrane configuration was also made. Configurations available were tubular, spiral-wound, flat-sheet and hollow fiber. Flat-sheet was rejected on the ground of high capital cost and complexities of membrane replacement. Hollow-fiber was rejected because they could not tolerate the high operating pressure. Both tubular and spiral-wound were considered to be appropriate for the duty. Both had their advantages and disadvantages. However, eventually, tubular configuration was selected because it could handle slurries and there was no requirement for pre-filtration.

3.4 Pilot Plant Trials

The next stage in the development of the membrane process involved pilot plant trials. The objectives of the trials were to :-

a) Obtain design and scale-up data and

b) Generate sufficient quantities of membrane processed liquor to carry out full down-stream processing and obtain product in the final form for customer assessment.

3.4.1 Design and Scale-up Data

The trials were designed to assess the following :-

a) Concentrations achievable for each product.

b) Assessment of cycle time for each product.

c) Most cost-effective way of achieving the operating pressure and cfv.

d) Closely linked to above was how best to configure the required membrane area.

e) Assessment of the number of batches of the same product which could be processed before a chemical cleaning was required.

f) Long-term fouling effects.

g) Optimisation of the cleaning procedure.

3.4.2 Down-stream Processing and Customer Assessments

Down-stream processing and customer assessment of the trial products led to the optimisation of the purification and concentration levels.

4. Full Scale Design

The essential requirements of the full scale plant were :-

a) The cycle time of the membrane process should be lower than the synthesis cycle time.

b) The membrane plant should have high availability.

Following the lab. scale and pilot plant trials a mathematical model was developed based on the gel layer model cited in literature[1] for the concentration step and assuming an average permeate flux for the diafiltration step. The predicted cycle time matched the achieved cycle time extremely well. The model was used to see the effect of various parameters (Feed flow, cross-flow velocity, effective pressure, average pressure, concentration factors etc.) on the cycle time of the process.

The cycle time of the membrane process had to be sufficiently low so that it did not become the bottle-neck item in the manufacturing chain. The cycle time requirement determined the membrane area requirement.

High availability of the plant was achieved by duplicating the critical plant items such as high pressure feed pumps, required membrane area etc.. For plant layout, consultations with the plant personnel led to several features which helped in the re-membraning process and made the maintenance easier by installing those plant items which needed a great deal of maintenance at accessible positions.

5. Rejection Mechanisms and Chemical Development

The membranes eventually chosen had a nominal MWCO well in excess of the molecular weight of the product species. This fact led to the speculation of the following rejection mechanisms :-

a) The effective molecular weight of the product species is greater than the MWCO of the membrane due to the aggregation of the product molecules.

b) The membrane material contains similar chemical groups as the product molecules. This can cause electrostatic repulsive forces between the membrane and the product molecules, thus giving good rejection of the product molecules.

c) The membrane, when initially treated with the product, develops a gel layer of the product molecules which itself acts as a secondary membrane layer for the rejection of the product molecules.

There is experimental evidence for all of the above mechanisms. Perhaps, the excellent rejections observed for the product molecules is a combination of all three mechanisms.

The membranes are not selective between the product molecules and impurities which have a similar molecular weight as the product molecules. Also, if the amount of impurities which do pass through the membranes is high then obviously the cycle time is elongated to reduce the amount of impurities to the required levels. These two facts obviated the need for synthesis process improvements to reduce the quantities of both types of impurities.

6. Operating Experience

The membrane process has been in operation for over two years and a great deal of operating experience has been gathered during this period. The main points to note are :-

a) Cycle times achieved are to design.

b) Membrane life has been in excess of 1.5 years with the cleaning procedure developed. Membranes are replaced when the cycle times of the membrane process exceed the synthesis cycle times.

c) Plant has run to the designed availability.

Any changes to the synthesis stages in terms of raw material change or addition of new raw materials are first assessed in the lab scale unit before implementation on the plant.

7. Concluding Remarks

The membrane process has given ZENECA considerable capacity, financial and environmental advantages over the traditional process.

The pay-back period based on raw material and yield savings alone was less than 6 months. However, this was overshadowed by the opportunities that the membrane process allowed in terms of telescoping various processes and improving product quality.

A rigorous development program together with the consultation and collaboration with the plant personnel and development chemists has meant the successful design and operation of the new process.

Nanofiltration membrane process is an exciting new separation technology for separation of high MW species from low MW species. Currently, the commercially available membranes are not selective between species which do not have large molecular weight differences. The capital and running costs of nanofiltration plants are high as well. Developments in enhancing permeate flux and membrane selectivity would be extremely beneficial for the fine chemicals industry.

References

1. R. Rautenbach, R. Albrecht, Membrane Processes, Wiley, Chichester, 1989.

MEMBRANE GAS ABSORPTION
A NEW TOOL IN SUSTAINABLE TECHNOLOGY DEVELOPMENT

Albert E. Jansen, Rob Klaassen, Paul H.M. Feron,
TNO Institute of Environmental and Energy Technology
Department of Chemical Engineering
P.O. Box 342
7300 AH Apeldoorn
The Netherlands

Abstract

The principle and engineering aspects of membrane gas absorption are discussed. The advantages of membrane absorbers compared to conventional absorption equipment, such as increased efficiency and flexibility, are due to the use small size hollow fibre membranes for contacting. Three example applications are discussed in which attention is also focused on the reuse of the absorbed chemical components.

1. INTRODUCTION

The necessity for the development of improved separation processes and more efficient equipment is of considerable importance during the transition of our society to a sustainable one. Recycling and reuse of material flows will become one of the dominant features in our industrial framework.

Process integration should be able to curtail emissions of hazardous substances and environmentally damaging components. Tighter environmental legislation calls for new equipment which is able to deal with removal of components over a wide range of concentration levels and a wide range of volume flows. This equipment needs to be implemented at various places within the process.

Membrane technology is an option rapidly gaining recognition for its flexibility and efficiency. Conventional membrane processes, such as reverse osmosis, ultrafiltration and microfiltration, have already been introduced successfully in various industries.

A novel application of polymer membranes is their operation as contactors. In this paper the attention is focused on the use of porous hydrophobic membranes for gas absorption duties. The resulting process is called membrane gas absorption. The principle and engineering aspects of membrane gas absorption are presented. Three applications are described in which the possible reuse of the absorbed component is highlighted.

The reuse of absobents can in many cases be combined with the regeneration of the absorption liquid. If this regeneration is efficient high purity products can be obtained. Recovery and reuse of components in the process industry can therefore contribute to a more sustainable economy.

2. PRINCIPLE OF MEMBRANE GAS ABSORPTION

The membrane forms a permeable barrier between a liquid and a gaseous phase. Gaseous pollutants diffuse through the pores and are absorbed by the liquid. The contact surface area between gas and liquid phase is maintained by the membrane and is independent of the gas and liquid flow rate. The selectivity of the partition is determined by the absorbent. Absorption in the liquid phase is determined either by physical partition or by a chemical reaction. Figure 1 shows the principle for a porous membrane.

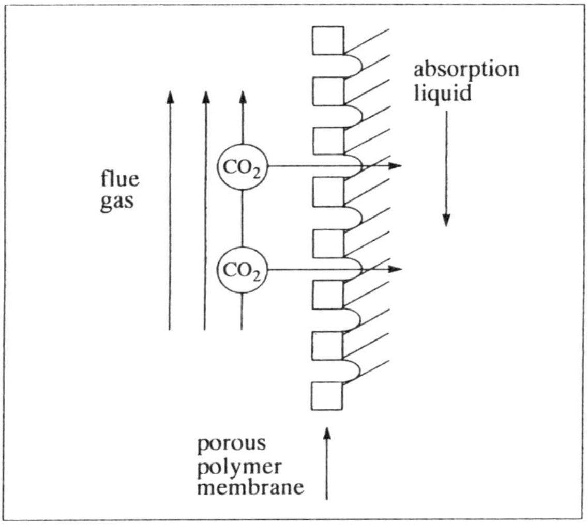

Figure 1 Principle of membrane gas absorption

Specifically by using hollow fibre membrane modules it is possible to develop a very efficient gas-liquid contactor. The majority of absorbents used in conventional gas absorption processes can also be used in membrane absorbers. For the operation of membrane gas absorbers it is essential that liquid phase and gaseous phase do not mix. This means that absorption liquids are not allowed to enter the pores. This is influenced by pore size, pressure difference across the membrane and the interaction of absorption liquid with the membrane material. These surface phenomena are described by the Laplace equation:

$$\Delta p = -2(\gamma/r)\cdot\cos\Theta \qquad (1)$$

in which:

Δp:	Pressure differential	$[N/m^2]$
γ:	Surface tension of liquid	$[N/m]$
r:	Pore radius	[m]
Θ:	Contact angle	

The membrane pores will not be wetted if the contact-angle is greater than 90° and the pressure differential is limited for a given pore size. Suitable membrane materials for aqueous absorption liquids are non-polar polymers, such as, polypropylene and teflon. As surface tension decreases with temperature break-through pressures also decreases with temperature.

A detailed description of mass transfer phenomena in membrane gas absorbers can be found in [1, 2, 3, 4].

3. ENGINEERING ASPECTS OF MEMBRANE GAS ABSORPTION

The use of gas absorption membranes has several advantages over conventional contacting equipment, like packed columns:

- the operation of the contact equipment is independent of gas- and liquid flow rates,
- no entrainment, flooding, channelling or foaming,
- compactness of the equipment through the use hollow fibre membranes.

The small dimensions of hollow fibres give rise to high specific areas, in excess of 1000 m^2/m^3. In principle this will lead to compact equipment with associated advantages in terms of cost and operation compared to conventional absorbers with specific contacting areas typically being 100 around m^2/m^3. It also leads to a reduction in absorber investment costs and a reduction in pumping power.

Membrane absorbers can be operated at liquid flow rates which are, unlike packed columns, optimal for the given process. This makes it possible to operate at extreme ratios (in excess of 1000) of the gas/liquid flow rates. These conditions cannot be realised in conventional contactors.

Current hollow fibre membrane modules are designed for filtration duties. Distribution of fluid around the fibres is ill-defined and leads to poor mass transfer. For further development of membrane gas absorption a new module design is absolutely imperative. The desired characteristics of such a module designed for contacting duties are:

- well-defined flow conditions on either side of the membrane,
- easy scale-up,
- countercurrent operation possible,
- high mass transfer,
- low pressure drop.

Figure 2 shows the basic design for a module intended for contacting duties [5]. One of the essential ingredients is a cross-flow mode of operation i.e. the feed flow is perpendicular to the fibres. The absorption liquid is fed through fibres. This design exhibits all of the desired qualities. An additional advantage over conventional membrane modules is the reduced frontal area as the feed flow is on the outside of the fibres. It is therefore not obstructed by the fibre potting. Scale-up can be achieved by stacking the membrane modules as shown in Figure 2.

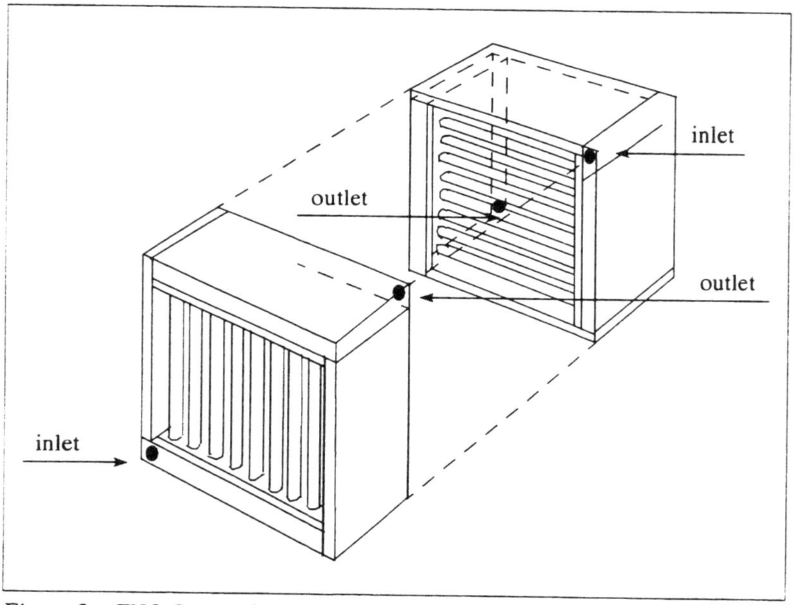

Figure 2 TNO Cross-Flow Membrane Absorber

4. APPLICATIONS

4.1 SO_2-removal from flue gas

A development program has been carried out for the removal of SO_2 from flue gas with membrane gas absorption. A pilot plant study during three production seasons on an industrial location (potato starch production) has been carried out with the aim of demonstrating the operation of the membrane absorber under on local conditions. Target was to reduce SO_2 emission at 100 m^3/h scale by at least 90%. At the same time it had to be determined whether the absorption liquid chosen could be reused in potato starch production replacing the sodiumbisulphite solution that is purchased at the moment.

The results obtained include:

- a high recovery of SO_2 (> 95%) from flue gas,
- the membranes guaranteed an easy and flexible operation,
- the membranes were not sensitive to fouling,
- reuse of the absorption liquid is possible.

First estimates of the costs of a membrane absorber indicate that a significant reduction in investment is possible compared to a conventional scrubber. A scale up for commercial application is foreseen in the beginning of 1996. The sulphite solution produced can also be used in other industries such as the food or paper industry.

Proposed alternative processes are absorption with sodium hydroxide solutions which form sulphate solutions after oxidation with the oxygen remaining in the flue gas. These sulphate solutions can be split in sulphuric acid and sodium hydroxide by membrane electrolysis. Capital costs together with power costs seem to be less then the sodium sulphate costs which makes the process economically feasible.

Yet another alternative is the crystallization of sodiumsulphate 10 aq. (Glaubersalt) at 10 °C after absorption at 50 °C.

4.2 CO_2-production from flue gas

In the Netherlands carbon dioxide enrichment in greenhouses is widely practised. Increasing the carbon dioxide concentration in the greenhouse from 350 ppm to 500 ppm results in a production increase of 25% for certain bulk crops. Carbon dioxide is currently generated by the combustion of natural gas which is wasteful as the resulting energy produced is not always utilised. The resulting flue gas may also contain impurities such as nitrous oxides and ethene which are detrimental to plant growth. Furthermore carbon dioxide requirement is high during summer when there is little or no energy requirement.

The current process is therefore not an energy-efficient process. Carbon dioxide can also be produced in a cogeneration plant with an absorption process. The waste heat from the cogeneration plant is used to regenerate the absorption liquid. The carbon dioxide thus produced is of high purity and can be used for enrichment in greenhouses without detrimental consequences for the plants.

The flow sheet of a membrane gas absorption process, shown in Figure 3, is similar to a conventional absorption process flow sheet.

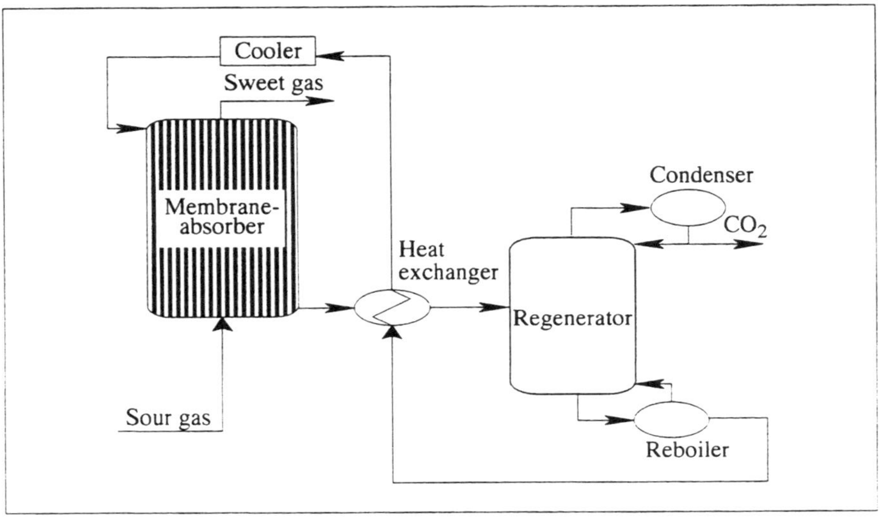

Figure 3 Flow sheet of the membrane gas absorption process for carbon dioxide removal

The flue gas containing carbon dioxide is countercurrently brought into contact with the absorption liquid in the membrane absorber. The rich absorption liquid is fed via a heat exchanger to the regenerator where carbon dioxide is released at high temperature.

The performance of a membrane absorber can be compared with the performance of a conventional packed column. It has been evaluated for a situation in which the flue gas at a temperature of 45 °C contains 3.5% CO_2. 80% Of the carbon dioxide is removed in the absorber. A mass transfer analysis indicates that the mass transfer resistance is dominated by the resistance in the liquid in both cases. The results of the comparison show that a membrane absorber can lead to a near ten-fold reduction in absorber size for this application [6].

Commercially available porous polyolefin membranes appear not to be stable for operation with the usual absorption liquid, i.e. aqueous solutions of monoethanolamine. In the course of time the liquid will seep through these membranes as a result of the surface tension of the organic liquid/water mixture. As a membrane material teflon is now a better choice because it is more hydrophobic and more inert towards aqueous monoethanolamine solutions [7]. Recently, new carbon dioxide absorption liquids have been discovered at TNO [8] which appear to be stable with polyolefin membranes. The use of these liquids, named CORAL, would enable the use of inexpensive membranes for carbon dioxide removal. A small scale test plant including regeneration by means of a conventional stripper has already shown good mass transfer results with these new absorption liquids. A further benefit of these liquids is there chemical stability compared to monoethanolamine.

4.3 H_2S removal

Due to its toxic character, the unpleasant smell and the acid rain problems H_2S emissions are in most cases no longer acceptable. With membrane gas absorption, in combination with an other technique H_2S can be converted into a valuable product.

The H_2S containing gas stream is led through the membrane absorber. In the membrane absorber H_2S is transferred very efficiently to an absorption liquid. The absorption liquid is then treated in the regeneration section and can be returned to the membrane absorber.

For the regeneration process there are two options: an electrochemical process or a crystallisation process.

In the electrochemical process the H_2S containing absorption liquid is converted into sulphur and a caustic solution. The high purity sulphur can be sold as product. The caustic solution is returned to the membrane absorber as absorption liquid.

In the crystallisation process sodium hydroxide is added in the regeneration unit. The Na_2S that will precipitate is removed as a product that can be sold. The caustic solution is returned to the membrane absorber.

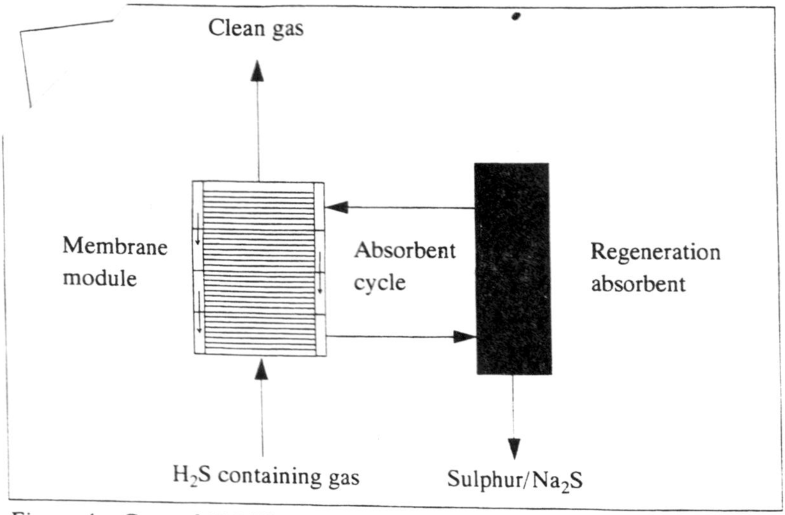

Figure 4 General H_2S Flowsheet

In summary a H_2S containing gasstream and electricity or sodiumhydroxide is entering the process and exiting the process are cleaned gas and sulphur or Na_2S as saleable products.

5. CONCLUSIONS

Membrane gas absorption is a novel process which in combination with suitable regeneration processes can result in the production of saleable products as well as prevention of environmental problems. The associated equipment is flexible and small compared to the conventional equipment. This paper has focused on the treatment of gasstreams containing acidic components. Beside these components development work is also focusing on removal of basic components (NH_3) and oxidative absorption liquids, e.g. for the removal of mercury from gasstreams.

6. REFERENCES

[1] Zhang, Q., Cussler, E.L.
'Microporous hollow fibers for gas absorption I. Mass transfer in the liquid'. Journal of Membrane Science, 23 (1985) 321-332.

[2] Zhang, Q., Cussler, E.L.
'Microporous hollow fibers for gas absorption II. Mass transfer across the membrane'. Journal of Membrane Science, 23 (1985) 333-345.

[3] Nii, S., Takeuchi, H.
'Removal of CO_2 by gas absorption across a polymeric membrane'.
Journal of Chemical Engineering of Japan, Vol. 25, No. 1, pp. 67-72, 1992.

[4] Nii, S., Takeuchi, H.
'Removal of CO_2 and/or SO_2 from gas streams by a membrane absorption method'.
Gas Separation & Purification, Vol. 8, No. 2, pp. 107-114, 1994.

[5] Meulen, ter, B. Ph.
'Transfer device for the transfer of matter and/or heat from one medium flow to another medium flow'.
US Patent nr. 5230796, European Patent nr. 0509031.

[6] Feron, P.H.M., Jansen, A.E.
'Capture of carbon dioxide using membrane gas absorption and reuse in the horticultural industry'.
Energy Convers. Mgmt. Vol. 36, No. 6-9, pp. 411-414, 1995.

[7] Falk-Pedersen, O., et. al.
'Separation of carbon dioxide from offshore gas turbine exhaust'.
Energy Convers. Mgmt., Vol. 36, No. 6-9, pp. 393-396, 1995.

[8] CORAL patent application filed 25 March 1994.

REACTORS

The Use of a Mass Transfer Efficient Bubble Column Reactor for Process Intensification in Selective Three Phase Catalytic Hydrogenations

J.M.Winterbottom, A.P. Boyes, S.Raymahasay, S.Sharma

School of Chemical Engineering, University of Birmingham, Edgbaston, Birmingham, B15 2TT, UK

ABSTRACT - Mass transfer efficiency and reaction kinetic studies for the Stirred Tank Reactor (STR) and the Cocurrent Downflow Contactor Reactor (CDCR) are compared. The specific reaction undertaken in both reactors is the selective hydrogenation of crotonaldehyde to n-butyraldehyde by using a 5% palladium on charcoal catalyst. Kinetic data obtained in the STR has indicated the reactor to be operating under diffusional resistances as shown by the apparent energy of activation, 20.8 kJ/mol and $X_{L\text{-}S}$ values of 13.6 %. Utilisation of the CDCR in swirl mode has shown the reactor's ability to allow the reaction to approach surface reaction rate control, as confirmed by an activation energy of 54.2 kJ/mol.

INTRODUCTION

Scale up of conventional stirred tank reactors (STR) to a pilot plant or industrial scale is a problem commonly encountered in the chemical and process industry. Small stirred tank reactors may prove efficient during laboratory testing, but scale up of such reactors will involve a compromise in an increase in operational and capital costs, with a reduction in reactor performance in terms of mass transfer efficiency.

The development of the Cocurrent Downflow Contactor Reactor (CDCR)[1] as a three phase catalytic reactor, has been shown to eliminate these problems due to its simple design and structure, thus allowing ease of scale up. Optimisation in design of the CDCR has lead to improvements in reactor performance, with a reduction in both operational and capital costs. Advantages of the CDCR over conventional three phase reactors are:

- 100 % gas utilisation
- High gas hold up values (≈ 40-60%)
- Formation of high interfacial areas (1000-6000 m^2/m^3, depending upon bubble size)
- Close approach to gas- liquid equilibrium in short contact times
- Maximum utilisation of the catalyst activity

The CDCR has also been utilised as a fixed bed reactor, therefore eliminating the loss of precious metal catalyst, which is likely to occur during the filtration stage of slurry reactions[2]. It is known, also , that under diffusional control metal catalyst can be leached from the support leading to unacceptable economic losses especially for precious metals, therefore kinetic control is very desirable. Further improvements to the reactor's performance include the addition of swirl (tangential flow) to the lower section of the column, thus further increasing the overall mass transfer efficiency of the reactor[3,4].

In the development of 'cleaner' processing technologies, this research is centred on the utilisation of the mass transfer efficient CDCR for selective hydrogenation reactions, with a comparison of reactor performance to the STR. Mass transfer coefficients and reaction kinetic data were evaluated for the selective hydrogenation of crotonaldehyde to n-butyraldehyde as a model reaction in the synthesis of a fine chemical intermediate. For all reactions within the STR and CDCR a highly selective palladium on charcoal catalyst was utilised.

EXPERIMENTAL

Catalytic Reaction Studies : Crotonaldehyde Hydrogenation

Aqueous solutions of crotonaldehyde (Aldrich, purity >90%, predominately trans) were hydrogenated in the presence of the palladium catalyst in a slurry system. Use of such a catalyst was expected to give rise to high selectivities of n-butyraldehyde, owing to selective hydrogenation of the olefin group.

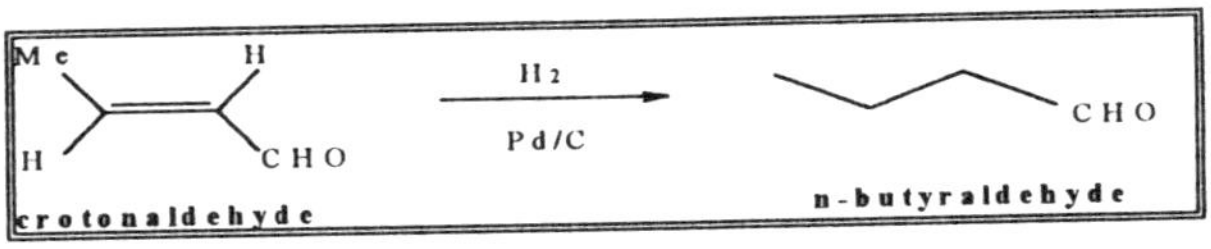

Reactant concentrations for mass transfer studies were maintained at 25.0 kg/m^3 and varied in the range 10.0-70.0 kg/m^3 in the determination of reaction orders with respect to substrate concentration as illustrated in Figure 2 (see later for discussion).

The catalyst employed was a 5% w/w Pd/C, Type 37 powder catalyst with a mean particle size of 19 μm and particle density of 2150 kg/m^3 (Johnson-Matthey Plc.). Pore diffusional problems commonly encountered in three phase reactions were eliminated, since the active metal had been deposited in the exterior surface of the support. The effectiveness factor of such a catalyst can therefore be assumed to be equal to one. Studies undertaken by Kenny and Sedricks[5] for crotonaldehyde hydrogenations to n-butyraldehyde by using a Pd/Al_2O_3 catalyst showed the effectiveness factor of their catalyst to be approximately 0.1.

Three Phase Catalytic Reactors

(I) Stirred Tank Reactor (STR)

Initial mass transfer and process efficiency studies were undertaken in the STR to obtain baseline information and modelling data for comparison to the CDCR. The STR employed was a conventional high pressure, stainless steel baffled autoclave incorporating a four bladed pitched stirrer. Reaction temperatures were obtained by use of a quick release heating unit. The STR of reactant volume 0.2 dm^3 was operated in the hydrogen partial pressure range 101.3-607.8 kPa and in the temperature range 298-343 K. A constant stirring speed of 1000 rpm was used, since at this value reaction rates were found to be independent of stirring speed.

(II) Cocurrent Downflow Contactor Reactor (CDCR)

The CDCR has evolved from a novel concept of contacting gas and liquid cocurrently, into a three phase catalytic reactor[1]. Gas (dispersed) and liquid (continuous) are introduced cocurrently at the top of the reactor, as shown in Figure 1. The interaction between gas and liquid produces a downward cocurrent flow, which passes through an orifice and entry zone creating a high velocity liquid stream. Entry of this stream into a fully flooded column creates a vigorously agitated gas-liquid dispersion in the top section of the column. This dispersion extends down the column length where coalescence of the bubbles takes place. The larger (coalesced) bubbles rise back up the column to be redispersed and broken up by the inlet stream. Under optimum operating conditions the downward liquid velocity is balanced with the bubble rise velocity. Such a procedure leads to the formation of a distinctive and sharp 'cut off' of the dispersion, thus eliminating any gas entrainment and results in 100% utilisation of the reactant gas. The CDCR was operated with a reactor volume of approximately 2.0 dm^3 and within the pressure and temperature ranges of 172.3-303.9 kPa and 288-343 K.

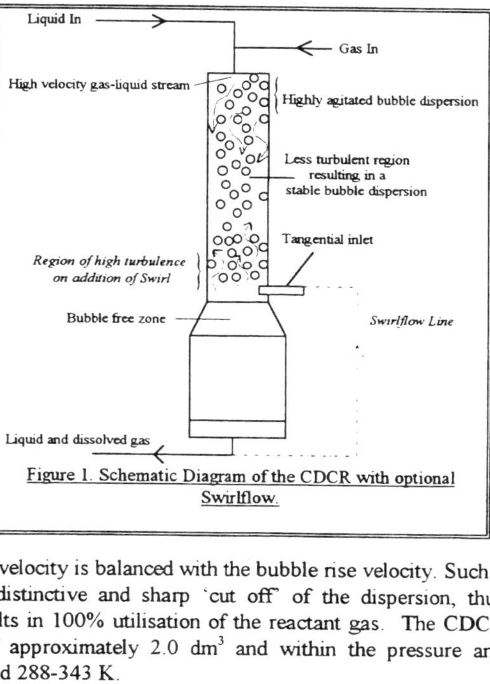

Figure 1. Schematic Diagram of the CDCR with optional Swirlflow.

Operation of the CDCR in swirl mode results in part of the slurry phase being recirculated back into the lower section of the column as shown in Figure 1, thus resulting in the break-up of the stable dispersion, creating greater turbulence and bubble shear and increasing the overall mass transfer efficiency of the reactor.

RESULTS

Mass Transfer and Reaction Kinetic Determination

Reaction orders with respect to crotonaldehyde concentration and hydrogen partial pressure were evaluated from Figure 2 and Figure 3, and were found to be zero order in crotonaldehyde and first order in hydrogen partial pressure. A first order relationship with respect to catalyst loading was also determined therefore allowing rate equations to be expressed as :

$$R_A' = k_2 P_{H_2}^{1.0} C_{Croton.}^{0} C_{Cat}^{1.0} \quad (1)$$

where k_2 = 2nd order rate constant

and k_r = pseudo first order rate constant = $k_2 C_{Cat}^{1.0}$ (2)

As the reaction showed first order characteristics, mass transfer coefficients were evaluated by using the classical film model for first order reactions[6,7] as shown by the following equation :

$$\frac{C*}{R_A'} = \frac{1}{k_L a} + \frac{\rho_p d_p}{6W}\left(\frac{1}{k_s} + \frac{1}{k_r}\right) \quad (3)$$

Liquid-solid mass transfer coefficients (k_s) were calculated from the Frössling Equation[8] where :

$$\text{Sherwood Number (Sh)} = \frac{k_s d_p}{D_A} = 2.0 + 0.76\, Re^{1/2}\, Sc^{1/3} \quad (4)$$

Calculations of k_s were based on liquid properties and not the catalyst particles, since in highly turbulent systems k_s can be assumed to be independent of particle size for small catalyst particles[6]. Equation 4 for non-swirl CDCR was applied to the highly turbulent top section of the column where gas and slurry first interact, through the orifice and 25-30 cm into the column, with a mean value of the Reynolds number (Re) calculated for these three regions of the CDCR. Sherwood number and hence k_s values, for swirl CDCR were calculated using equation (4) with the value of the tangential flow incorporated into the equation. Within the STR, k_s values were based on the mean value of the Sherwood number at the tip of the impeller, equation (4) and at the wall of the reactor, where the Sherwood number for particles in the latter case was represented by equation (5) :

$$Sh = 2 = \frac{k_s d_p}{D_A} \quad (5)$$

Energies of activation were also determined from the Arrhenius relationship, as a means of determining the controlling reaction regime within both the STR and CDCR :

$$\text{Where } k = A\exp^{\left(-\frac{E_A}{RT}\right)} \quad (6)$$

Mass transfer coefficients and energies of activation determined for the STR and CDCR are presented in Table 1. Determination of k_La from equation (3), showed comparable high values for all three reactor types, indicating negligible resistances to gas-liquid mass transfer.

Liquid-solid mass transfer coefficients (k_s) within the CDCR (non-swirl and swirl) were greater in value as compared to the STR throughout most of the temperature range. However, addition of swirl resulted in an increase in value for k_s, indicating greater mass transfer efficiency for a small increase in energy input to the reactor system. Liquid-solid mass transfer resistances, $X_{L\text{-}S}$, within the STR were still prominent and confirmed by the low apparent energy of activation, E_A of 20.8 kJ/mol.

Apparent energies of activation obtained from ln R_A' vs 1/T plots for the non-swirl CDCR and swirl CDCR were respectively in the range 27.0-37.7 kJ/mol and 34.4-40.5 kJ/mol; these indicate close approach to surface reaction rate control. Energies of activation obtained by calculating k_r using equation (3) in conjunction with equation (6) gave respectively 49.7 and 54.2 kJ/mol for non-swirl and swirl modes. The latter values are deemed to be true activation energies, associated with the surface chemical reaction. The apparent E_A values for the swirl CDCR are quite close to these and demonstrate ability of swirl flow to reduce mass transfer resistances to a low level. However, $X_{L\text{-}S}$ values determined for non-swirl and swirl CDCR, indicated that liquid-solid mass transfer resistances were still prominent at high temperatures.

Table 1. Mass Transfer Coefficients and Energies of Activation for Crotonaldehyde Hydrogenation

Reactor configuration	k_La (s^{-1})	k_S x10^2 (m/s)	k_r x10^2 (m/s)	$X_{L\text{-}S}$ (%)	E_A (kJ/mol)
STR	2.1 - 4.4	4.5 - 9.5	0.62 - 1.5	12.1 - 13.6	20.8
CDCR (Non-swirl)	2.7 - 3.3	6.3 - 14.6	0.4 - 6.1	5.5 - 29.7	49.7
CDCR (Swirl)	1.7 - 6.7	9.3 - 21.5	0.5- 7.3	5.0 - 26.6	54.2

Product analysis by gas chromatography confirmed high selectivities (100%) towards n-butyraldehyde within both the STR and CDCR, thus representing clean synthesis with zero side reactions. Figure 4 illustrates the production of n-butyraldehyde as a function of time within non-swirl and swirl CDCR. Application of swirl flow to the reactor system can be seen to reduce reaction times by a factor of two, in conjunction with higher gas hold up values ($\approx$60%) and enhanced values of k_s being obtained.

CONCLUSION.

Comparable rates of gas-liquid mass transfer were obtained in the STR and CDCR. Liquid-solid mass transfer coefficients were found to be greater in the CDCR than in the STR, this was obtained with the CDCR operating at a volume ten times greater than the STR. Calculation of the power input per unit volume of liquid, has shown an increase in energy requirement of 25 times if the STR were to be scaled up to the volume of the CDCR. For a much smaller increase in energy input to the CDCR, swirl flow has been shown to further optimise the reactor performance with complete utilisation of the catalyst activity and an increase in rates of hydrogenation. Future studies on crotonaldehyde hydrogenation will involve the use of modified platinum catalysts, with the ability to give selective production of the unsaturated alcohol, crotyl alcohol.

ACKNOWLEDGEMENTS.

The authors are grateful to EPSRC and ICI Katalco for the provision of a research grant and to EPSRC for the provision of a studentship (S.Sharma) and to Johnson Matthey Plc. for loan of the precious metal catalysts.

NOMENCLATURE.

ap	Catalyst interfacial area	m^2/m^3
C*	Equilibrium concentration of gas at the gas-liquid interface	mol/m^3
C_{Cat}	Concentration of catalyst.	mol/m^3
C_{Croton}	Concentration of crotonaldehyde	mol/m^3
dp	Diameter of catalyst particle	m
d	Diameter of T-piece, orifice or reactor column	m
D_A	Diffusivity of hydrogen	m^2/s
E_A	Energy of activation	kJ/mol
k_La	Volumetric gas-liquid mass transfer coefficient	s^{-1}
k_s	Liquid solid mass transfer coefficient	m/s
kr	Surface reaction rate constant (pseudo first order rate constant)	m/s
P_{H2}	Hydrogen pressure	kPa
R_A'	Reaction rate per unit volume of clear liquid/slurry	mol/m^3s
Re	Reynolds number = $\rho_L U_L d / \mu_L$	-----
Sc	Schmidt number = $\mu_L / \rho_L D_A$	-----
Sh	Sherwood number	-----
T	Temperature	K
ρL	Density of the liquid	kg/m^3
ρp	Density of the catalyst particle	kg/m^3
μL	Viscosity of the liquid	kg/ms^2
UL	Liquid velocity	m/s
X_{L-S}	Liquid solid mass transfer resistances	%

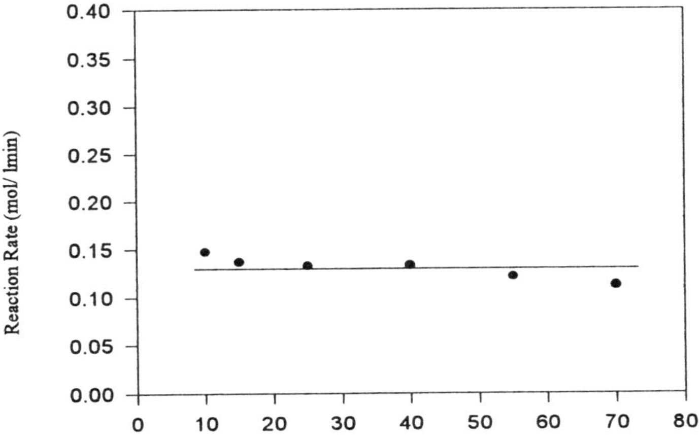

Conc $_{Croton.}$ kg / m^3

Figure 2. Rate of hydrogenation versus Crotonaldehyde Concentration. Data obtained at 600 kPa , 298 K and catalyst loading 1.0 kg/m^3 in the STR

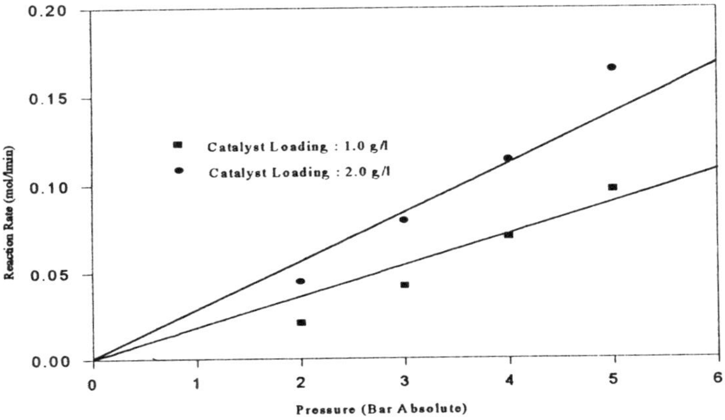

Figure 3. Reaction Rate versus Increasing Hydrogen Partial Pressure. Data obtained at 298 K, 1000 rpm, Conc$_{Croton}$: 25.0 kg/m^3 in the STR

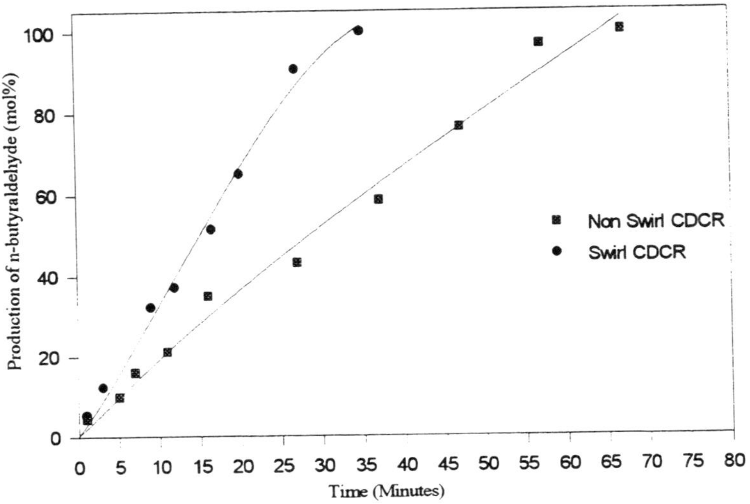

Figure 4. Production of n-butyraldehyde as a Function of Reaction Time in the CDCR (non-swirl and swirl). Data obtained at 303 K and 303.9 kPa

REFERENCES.

(1) Boyes, A.P. and Ellis, S.R.M., (1976), Gas Liquid Contactor, UK Patent No 1, 596, 738A.

(2) Boyes, A.P., Chugtai, A., Khan, Z., Raymahasay, S., Sulidis, A., Winterbottom, J.M., (1995), The Cocurrent Downflow Contactor (CDC) as a Fixed Bed and Slurry Reactor for Catalytic Hydrogenation. Journal of Chemical Technology & Biotechnology, **64**, 55-65

(3) Khan, Z., Boyes, A.P., Raymahasay, S., Winterbottom, J.M., (1994), The Cocurrent Downflow Contactor Reactor (CDCR): Catalytic Hydrogenation of Model Three Phase Reaction Systems In Slurry Operation With and Without Tangential Flow., IChemE Research Event, 652-655

(4) Boyes, A.P., (1989), Swirlflow CDC, US. Patent 4834343.

(5) Kenny, C.N., and Sedricks, W., (1972), Effectiveness Factors in a Three Phase Slurry Reactor : the Reduction of Crotonaldehyde over a Palladium Catalyst, Chem. Engng. Sci., **27**, 2029

(6) Ramachandran, P.A. and Chaudhari, R.V., (1983), Three Phase Catalytic Reactors., Gordon and Breach Science Publications

(7) Doraiswamy, L.K. and Sharma, M.M., (1984), Heterogeneous Reactions., Volume 2, Wiley Interscience

(8) Satterfield, C.N., (1970), Mass Transfer In Heterogeneous catalysis., M.I.T. Press, Massachusetts

SELECTIVE LIQUID-PHASE HYDROGENATION OF STYRENE AND 1-OCTENE MIXTURES IN A MONOLITHIC REACTOR

A. STANKIEWICZ, H. A. SMITS, W. Ch. GLASZ
DSM Research, P.O. Box 18, 6160 MD Geleen, The Netherlands

and

J. A. MOULIJN
Delft University of Technology, Department of Chemical Engineering,
Julianalaan 136, 2628 BL Delft, The Netherlands

Abstract

In the paper the results of the experimental studies on liquid-phase hydrogenation of 1-octene/styrene mixtures in toluene, carried out in a monolith reactor, are presented. The relative importance of the mass transport limitations and uneven liquid distribution in the overall reactor performance are discussed. The investigated reactor type appears to be superior to conventional designs commercially applied to this kind of hydrogenation processes, such as trickle-bed of packed bubble column reactors.

1. Introduction.

Monolith reactors present a very promising tool for improving the yields/selectivities of chemical processes. Monoliths are already commonly used for cleaning off-gases (e.g. as car exhaust catalysts). In recent years, the use of monolithic catalysts for improvment of yields and selectivities in gas-liquid processes has attracted considerable attention and led to the first commercial-scale application (hydrogenation of 2-ethyl antraquinone in the AKZO Nobel hydrogen peroxide technology).

The most important advantages of monolithic reactors operating in the so-called Taylor (segmented) flow hydrodynamic regime (Figure 1), are [1, 2]:

- very short diffusion path inside the catalyst;
- high gas-liquid and liquid-solid mass transfer coefficients;
- high contact areas per unit reactor volume (usually over 2000 m^2/m^3);
- easy scale-up, thanks to reproducible conditions in individual channels of monolithic catalysts;
- very low pressure drop, compared to conventional packed-bed systems.

It can be expected that in view of the above features of monolithic catalysts using them instead of conventional trickle-bed reactors should bring considerable profits in numerous commercial processes, especially those processes, in which activity and/or selectivity are controlled by the mass transfer. Among those processes hydrogenations and oxidations are the most important ones. Recently, a comparative analysis of a monolith and a trickle-bed reactor for a three-phase hydrogenation has been presented by Edvinsson and Cybulski [3].

In the present study the competitive hydrogenation reactions of styrene and 1-octene in toluene over a monolithic Pd catalyst are investigated. The above reactions are shown schematically in Figure 2. They may be considered as quite representative for a broader group of reactions taking place e.g. in hydrotreating processes of the naphta cracking cuts.

2. Experimental.

The experiments were carried out in the bench-scale monolith setup presented in Figure 3. The setup was operated in the batch mode (feed-batch with respect to H_2). Hydrogen and the investigated hydrocarbon mixture were pumped through the reactor, in which 30 cm long monolithic Pd catalyst was fixed. After passing the reactor the gas-liquid mixture was directed to the separator, from which both phases were recycled to the reactor, gas passing through the condensor, in order to separate the remaining small amounts of hydrocarbons. Fresh hydrogen was fed in quantity sufficient to keep the gas volumetric flow rate on the constant level. Samples of the hydrogenated liquid phase were taken at regular time intervals from the by-pass line of the liquid pump.

The above bench-scale setup was operated at temperatures of 40 - 90°C and under pressures 6 - 15 bara. The liquid-to-gas ratio and superficial velocity were varied between 1:1 - 6:1 litre/litre and 0.050 - 0.125 m/s, respectively.

The monolithic catalyst used in the experiments was a commercial Al_2O_3 washcoated monolith containing 400 square-shaped channels per square inch. The thickness of the catalytic layer was ca. 15 μm. Before starting the experiments the catalyst was reduced *in situ*.

The samples taken during the experiments were analyzed off-line by gas chromatography, with cumene being used as internal standard.

3. Discussion of results.

Figure 4 shows typical concentrations vs. time curves for hydrogenation of 1-octene/styrene mixture in toluene. Concentrations are plotted here as a percentage of the total quantity of reactants. The experiment presented in Fig. 4 was carried out at 50°C and under 10 bara, with liquid-to-gas ratio of 1:1 and liquid superficial velocity of 0.10 m/s.

Figure 5, in turn, shows the selectivitiy of preserving octenes (as a percentage of the initial amount) and the selectivity of formation of internal octenes (as a percentage of the initial amount of 1-octene) as functions of styrene conversion. Lines corresponding to both selectivities are marked by S(1) and S(2), respectively.

The above figures clearly show that styrene is preferably hydrogenated to ethylbenzene, while 1-octene is isomerized to internal olefines. This isomerization to internal octenes (shown as the fraction of total n-C8 pool) is presented in Figure 6. Only when styrene is depleted octenes get converted to n-octane at an increased rate. Adsorption of styrene appears to be much stronger than the competitive adsorption of octenes. The hydrogenations of aromatics to form alkylocyclohexanes can be assumed negligible, since it is always (much) less than 0.2%.

The rate constants for individual reactions were estimated assuming the Langmuir-Hinshelwood-Hougen-Watson kinetics. It appeared that liquid velocity (flow rate) had a significant effect upon the reaction rate constants, as it is shown in Figure 7. The rate constants presented in the figure correspond to the following reaction steps:

k_1 - 1-octene to octane;
k_2 - 1-octene to n-octene;
k_3 - n-octene to octane;
k_4 - styrene to ethylbenzene.

The observed increase of the reaction rates with liquid velocity indicates that the mass transfer limitations play a considerable part in the process under consideration. The internal mass transfer characteristics were evaluated by means of Weisz-Prater criterion, formulated as:

$$\Phi = \eta\varphi^2 = \frac{r_{obs}\rho_{cat}L^2}{D_{eff}c_{i,surf}} \ll \frac{2}{n+1}$$

where η is the effectiveness factor of the catalyst;
φ is the Thiele modulus;
r_{obs} is the observed reaction rate;
ρ_{cat} is the catalyst (washcoat) density;
L is the characteristic diffusion length (washcoat thickness in this case);
D_{eff} is effective diffusion coefficient in the catalyst pores;
$c_{i,surf}$ is the surface concentration of species "i";
n is the order of the reaction.

The evaluation showed that the internal limitation was negligible. This can be seen in Figure 8, where the Weisz-Prater moduli are presented as functions of the washcoat thickness. Clearly, for the thickness of 15 μm the values of Weisz-Prater moduli are small enough to neglect the internal mass transfer as the limiting step.

On the other hand, the possible presence of the external mass transfer limitation was checked by the Carberry number criterion:

$$Ca = \frac{r_{obs}}{k_L a \cdot c_{i,bulk}} \ll 0.05$$

where k_L is the mass transfer coefficient;
a is the interfacial area;
$c_{i,bulk}$ is the liquid-phase concentration of species "i".

For neither of three reactants involved (hydrogen, styrene, 1-octene) did the Carberry number fulfill the above criterion, assuming values between 0.08 and 1.24. This is a strong indication of the external mass transfer limitations appearing in the reactor under conditions investigated in the present study.

Another effect that can potentially play a role in the observed behaviour of the monolith reactor is the hydrodynamic factor, resulting for instance from the non-ideal distribution of the liquid among the individual channels. So far, we have assumed the Taylor flow in all the channels. However, other flow regimes, e.g. aerosol formation (with ver poor catalyst wetting) might also arise in a part of the channels. An indirect proof of the reaction course being influenced by hydrodynamic phenomena is the fact that, contrary to what could be expected from the uniform Taylor flow, the observed dependence of conversion on pressure was rather weak and the decreased liquid fraction led to lower conversions. If hydrogen transfer across the liquid film were the only rate limiting step during the entire test run, the reaction rates would be linearly proportional to pressure. The picture that we got, combined with the fact that reaction rates are almost linearly proportional to the velocity, give rise to the assumption that catalyst wetting was incomplete and/or non-uniform. The fact that we are likely to be dealing with liquid mal-distribution presents both a good and a bad niews. On the one hand it is the problem that needs to be solved (especially when coming to a larger-scale unit), on the other hand, however, it shows that quite considerable improvement in activity per unit reactor volume can be expected when liquid distribution is improved.

In order to see how our most likely imperfect reactor performs in comparison with other reactor types we started the search for some literature data regarding the styrene hydrogenation in more conventional reactor systems. Unfortunately, the literature in this field is very scarce and the only data useful for comparison could be extracted from the paper by Mochizuki and Matsui [4], who investigated the hydrogenation of styrene and phenylacetylene in a fixed-bed co-current upflow reactor filled with a Pd catalyst. Rough evaluation showed that at 50°C and under pressure of 6 bara, our hydrogenation rate of styrene (per kg of the catalyst) was approximately one order of magnitude higher than that reported in [4]. Bearing in mind all the imperfections of our monolith as well as the fact that our catalyst was not subject to any optimization, the above demonstrates clearly the superiority of the monolith reactor over the presently used conventional reactor types.

4. Conclusions.

The simultaneous hydrogenation of styrene/1-octene mixtures in a monolith Pd catalyst showed that styrene is preferably hydrogenated to ethylbenzene, while 1-octene is isomerized to internal olefins. Only when styrene is depleted octenes are converted to n-octane at increased rates.

It has been observed that the gas-liquid mass transfer limitations occured under conditions investigated. Furthermore, the results indicated possible non-uniform distribution of the liquid in the monolith channels. Despite these facts the rates achieved in the monolith reactor appeared to be higher by a factor of ca. 10 than the rates reported in the literature for an up-flow packed column reactor.

With improved liquid distribution and optimized catalyst as well as process conditions, further improvement of reactor performance is expected.

Our experiments proved that in case of liquid phase hydrogenations of unsaturated hydrocarbons the monolith reactors may present an attractive commercial alternative to conventional reactor designs such as packed bubble columns or trickle-bed units.

5. References.

1. Irandoust S., Andersson B., Catal. Rev. Sci. Eng., **30**, 341 (1988).
2. Cybulski A., Moulijn J.A., Catal. Rev. Sci. Eng., **36**, 179 (1994).
3. Edvinsson R. K., Cybulski A., Chem. Eng. Sci., **49**, 5653 (1994).
4. Mochizuki S., Matsui T., AIChE J., **25** , 904 (1976).

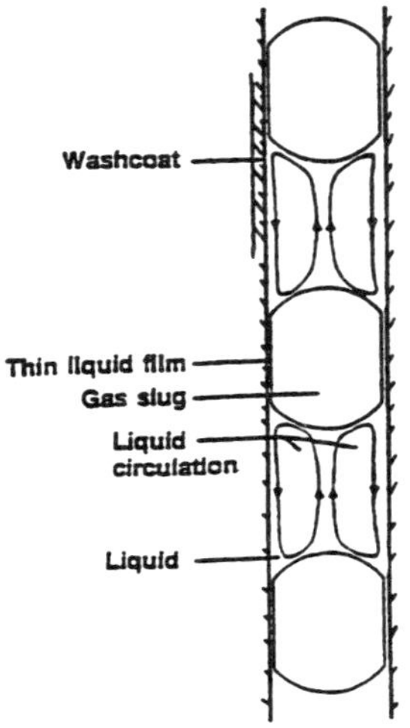

Fig. 1:
Taylor flow in a monolith channel

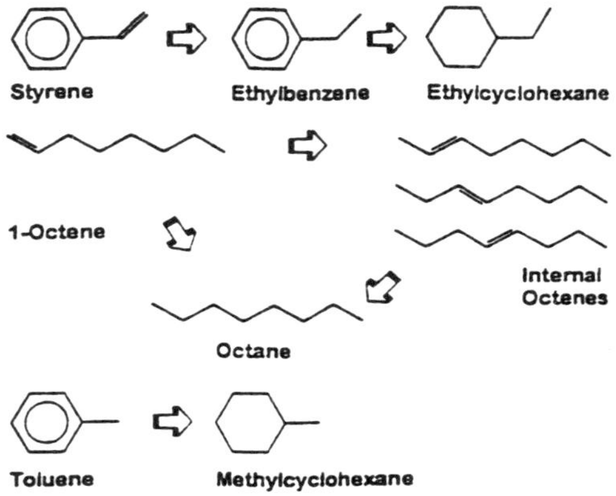

Fig. 2:
Scheme of reactions investigated

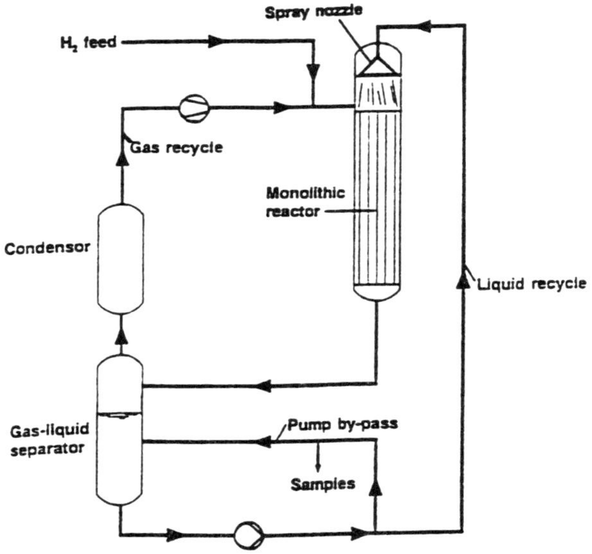

Fig. 3:
Scheme of the experimental setup.

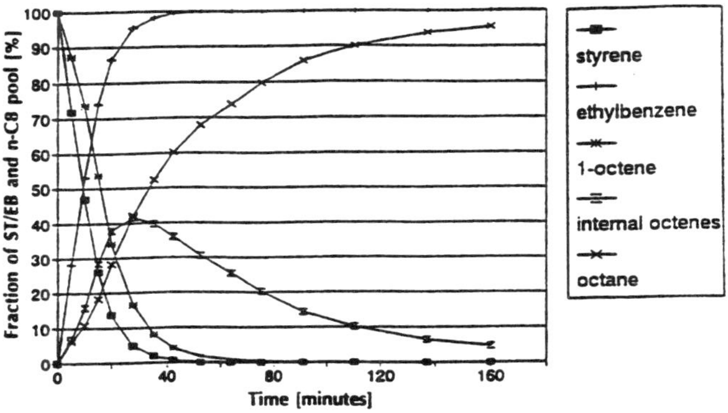

Fig. 4:
Hydrogenation of 1-octene and styrene in the monolith reactor

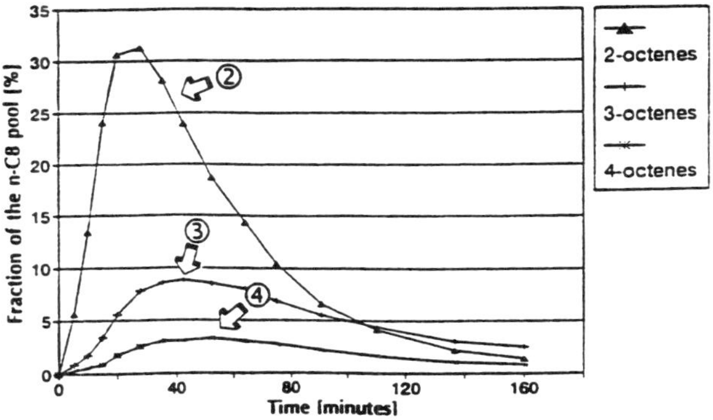

Fig. 5:
Isomerization to internal octenes in the monolith reactor

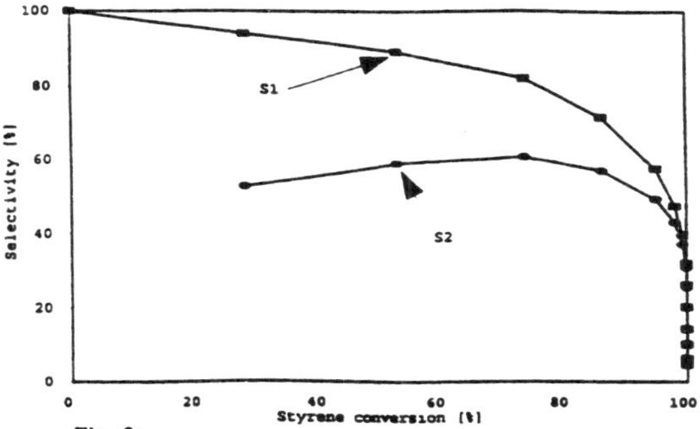

Fig. 6:
Selectivities of preserving octenes and of internal octenes formation as functions of styrene conversion.

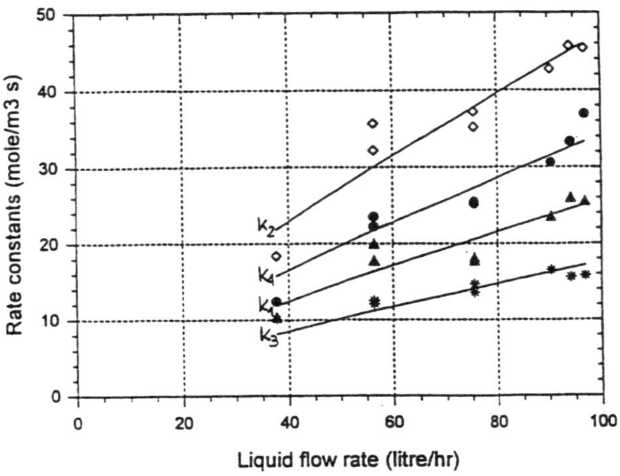

Fig. 7:
Reaction rate constants as a function of liquid flow rate.

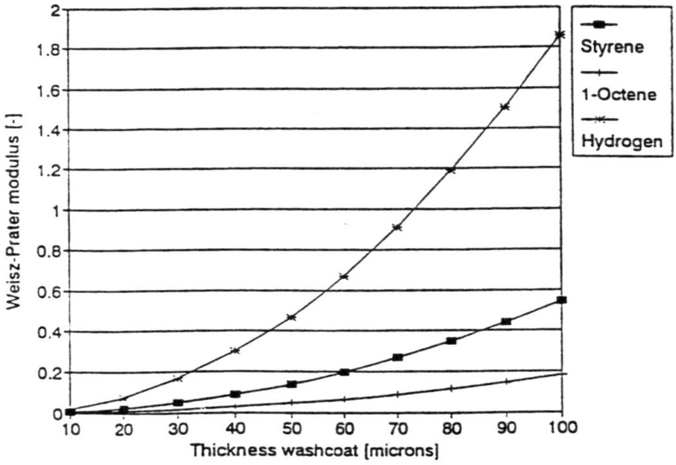

Fig. 8:
Weisz-Prater criterion as a function of washcoat thickness.

PROCESS INTENSIFICATION: SPINNING DISC POLYMERISER FOR THE MANUFACTURE OF POLYSTYRENE

K.V.K. Boodhoo, R.J.Jachuck* & C. Ramshaw
Department of Chemical & Process Engineering
University of Newcastle upon Tyne, UK
(For Correspondence)*

1. Introduction

The spinning disc concept is an attempt to apply process intensification methods within the field of heat and mass transfer. The technology operates by the use of high gravity fields created by rotation of a disc surface causing the reacting fluid to flow in the form of thin wavy films. Such thin films have been shown to significantly improve the heat transfer rates. Studies carried out more recently under the Joule 1 programme at the University of Newcastle Upon Tyne have also shown that the inclusion of irregularities in the form of grooves, for example, on the disc surface helps to enhance the heat transfer performance by inducing the formation of a large number of small waves which create instabilities in the film [1]. The high heat and mass transfer characteristics achieved on a rotating disc suggests that it could be used as a reacting surface for systems which are heat and mass transfer limited. This applies to most polymerisation reactions, where the high viscosity build up during the reaction inhibits good mixing and hence restricts mass transfer between the reactants. The development of a spinning disc reactor under the Joule 2 programme at this university has enabled tests to be carried out to investigate its performance in relation to the polymerisation of unsaturated polyester [2]. The work on styrene is currently being carried out under the Joule 2 supplement programme.

2. Experimental facility

The apparatus used for the experimental work consisted of a batch reactor and the spinning disc reactor. The batch reactor for polyester production was in the form of a 250ml glass conical flask, with magnetic stirrer and hot plate (Figure 1); this was later improved to one having a water circulator surrounding the vessel as a constant temperature bath for the polystyrene production .

The spinning disc equipment, as shown in Figure 2, consisted of five main sections: the spinning disc surface, the slip ring assembly incorporating thermocouples, the radiant heaters, the prepolymer feed distributors and the cooling coils. The rotating disc was made of brass and had concentric grooves at varying radial distances to provide better mixing and increased surface area. The disc surface temperature was measured using thermocouples embedded at chosen locations and connected to the data acquisition computer.

3. Experimental procedure

The polymerisation reaction of styrene consisted of two stages: (1) prepolymer stage (2) spinning disc stage.

Prepolymer stage

200 ml of styrene monomer was preheated to the operating temperature of 88°C in the batch reactor. This temperature was maintained throughout the reaction by the water bath and was regulated by a digital control system. The vessel had a magnetic stirrer and was connected to a condenser at the top. Benzoyl peroxide initiator (3g) was then added to the monomer through a stopped arm of the vessel. When the desired conversion of the prepolymer was obtained, the melt was poured down onto the spinning disc surface via the feed distributor.

Spinning disc stage

The spinning disc surface was heated to 88°C by an electric heater and its rotational speed was set at about 350-400 rpm. The disc was allowed to rotate for about 4-5 minutes to ensure that almost all of the feed introduced had sufficient time to flow down the distributor tubes and come into contact with the disc surface. After passing over the disc surface, the polymer melt was thrown off the disc surface into the collecting trough which was water-cooled. Samples of the spinning disc product were taken immediately after the disc stopped rotating and the heater was switched off. Samples were then analysed to determine the conversion achieved and the molecular weight.

4. Analysis of samples

The conversion was calculated by weighing the samples before and after polystyrene was precipitated from the monomer and polymer mixture. Methanol was used as precipitating agent; the monomer styrene remained in solution, leaving polystyrene as a white precipitate which was washed with more methanol and dried at 65-70°C to constant weight.

The molecular weight analysis was also carried out.

5. Results and Discussion

Conversion

Results of the tests which were carried out to investigate the change in conversion of the polymer melt as it passed from the batch reactor onto the spinning disc surface have been displayed in Table 1. These results could be more clearly interpreted by analysing the time-saving achieved on the spinning disc polymeriser compared to a conventional batch reactor. The savings in time which have been obtained for different feeds into the spinning disc reactor have been more explicitly shown in graphical form (Figure 3). The conversion range over which the performance of the spinning disc reactor has been investigated has been restricted to a maximum value of about 60% as the high viscosity attained beyond this point in the batch mass polymerisation process makes the flow of the reacting medium rather difficult when pouring onto the rotating surface. As seen in Table 1 and Figure 3, the saving in time is roughly doubled from 7 minutes to 15 minutes as the feed conversion increases from 0.12 to 0.28, indicating that the performance of the spinning disc reactor with respect to the conventional batch reactor gets better as the conversion goes up. The rate of reaction achieved on the spinning disc is therefore higher than that in the batch reactor, giving a quite significant reduction in reaction times.

TABLE 1: Changes in conversion and time savings in Spinning Disc Reactor (SDR)

Time of feeding into SDR (mins)	Feed conversion	Product conversion	Equivalent batch time (mins) (Time savings by using SDR)
15	0.12	0.22	22
20	0.18	0.32	34
25	0.24	0.34	35
30	0.28	0.43	45
35	0.33	0.53	52
40	0.37	0.57	57

Disc Temperature : 88°C
Disc Speed : 350-400 rpm

6. Merits of the Proposed Technology

The spinning disc technology is expected to make the polymerisation process intrinsically safer and more energy efficient. Previous results from the work done on polyester have established that a cascade of 9 discs would give the performance of 6 hours batch reaction; this would not only have a major impact on production and profitability but the process intrinsic safety would also be largely improved. Furthermore, on an industrial scale, polymer products of narrow molecular weight distributions are expected from reaction on the spinning disc, hence better quality polymers.

7. Conclusions

The research work on the spinning disc reactor done so far has established that the reactor has strong potential to be used as a polymeriser due to its enhanced heat and mass transfer characteristics. Significant reductions in the reaction times and tighter control of the molecular weight distribution of the polymer may be possible on the spinning disc polymeriser. Finally, a spinning disc polymeriser could significantly improve the intrinsic safety of the polymerisation process. However, before any major conclusions can be drawn, detailed studies have to be carried out to understand the effect of system variables such as disc temperature, disc rotational speed and the prepolymer flow rate on the polymerisation process of styrene in the spinning disc reactor.

References

(1) Jachuck, R.J., & Ramshaw, C., Heat Transfer Characteristics of Tailored Rotating Surfaces, Eurotherm Seminar No.26, 1992.

(2) Jachuck, R.J., & Ramshaw, C., Process Intensification: Spinning Disc Polymeriser, ICHEME Research Event / First European Conference, 1995

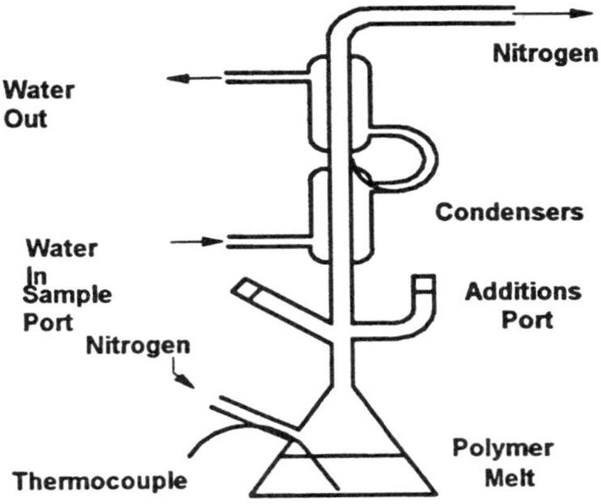

Figure 1 Prepolymer Stage Batch Reactor

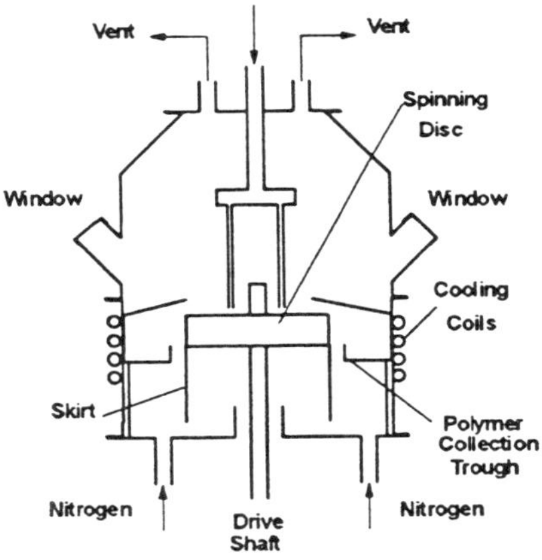

Figure 2 Spinning Disc

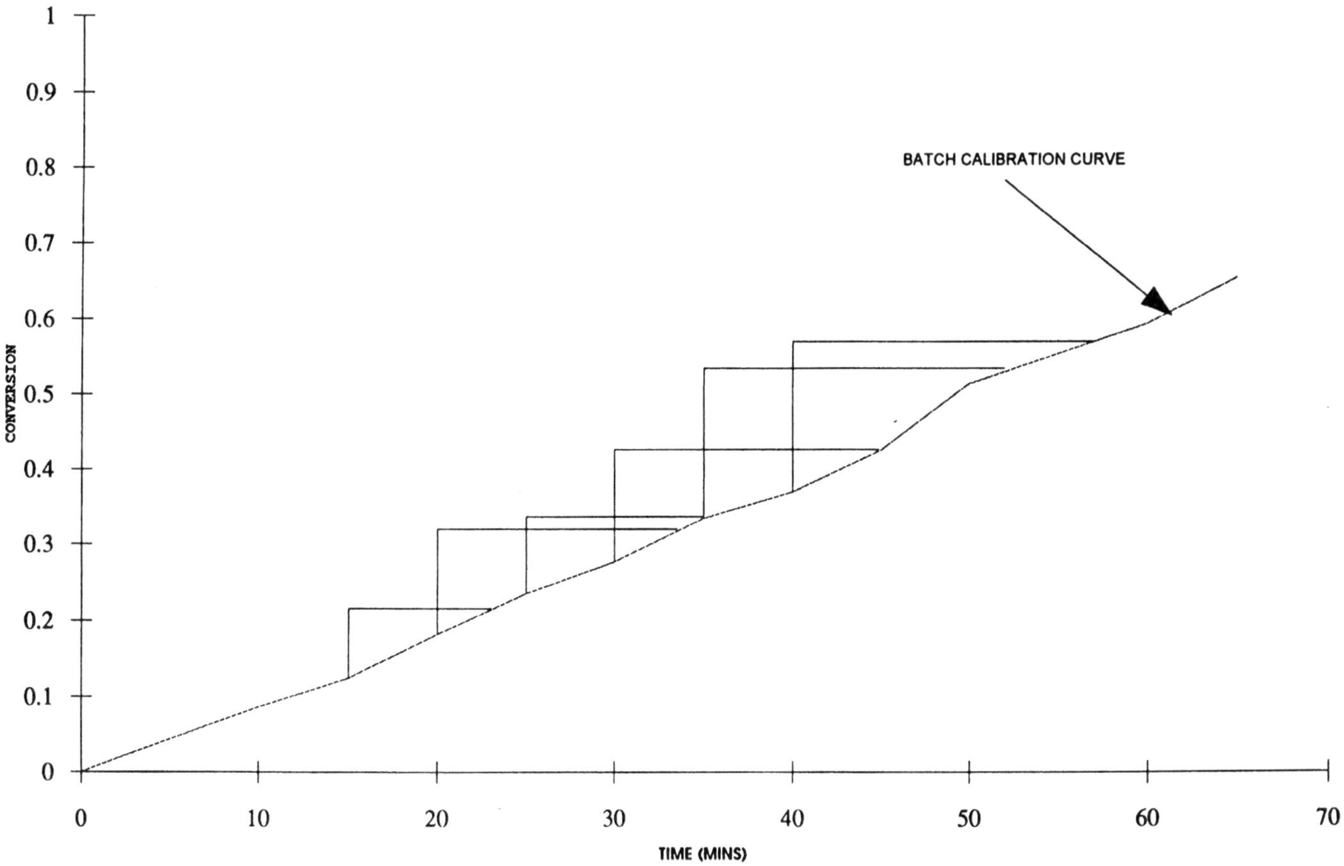

FIGURE 3. PERFORMANCE OF THE SPINNING DISC REACTOR RELATIVE TO BATCH REACTOR

THE JET -IMPINGEMENT REACTOR, A NEW HIGH-INTENSITY REACTOR FOR LIQUID-LIQUID REACTION PROCESSES

Dr. Edward G. Hauptmann
John M. Rae
Dr. Alfred A. Guenkel

NORAM Engineering and Constructors Ltd.
Vancouver, B.C., CANADA

SUMMARY

A novel high-intensity reactor system with no moving parts has been developed and is being used to great advantage in the production of mononitrobenzene (MNB). High velocity jets within the reactor are made to impinge on each other, generating a large interfacial area between phases, with resultant accelerated reaction rates and improvements in process efficiencies. The benefits of the jet-impingement reactor principle may be obtained in many other heterogeneous liquid/liquid systems.

INTRODUCTION

Mass transfer to interfacial reaction zones is known to be an important limiting factor in many heterogeneous liquid/liquid reactions [1]. Coalescence and redispersion of the reacting species complicate the complete understanding of overall reaction rates and conversion in such systems.

Nevertheless, the benefit of continuously creating fresh surface area for reacting species is obvious, and has led to well-established reactor designs based on agitation in stirred tanks, blending through static mixers, spray systems, or other such means [2].

In many cases, the dispersion of the reactants in a stirred vessel can be surprisingly slow, for example taking 10-20 minutes to disperse an organic into an aqueous phase [3], with generation of new surface area restricted mainly to regions near the agitator. Static mixers on the other hand perform adequately for blending components, but most designs have a limited capability to break up droplets of the reactant in the dispersed phase. Sprayed systems can generate small drops, but require relatively large amounts of power.

A novel method has been devised to provide high-intensity shearing action between liquid/liquid reactants, thereby generating abundant fresh interfacial area and continuously redispersing the reacting species. This has led to the development of the jet-impingement reactor, the main features of which are described in the following section. This reactor concept has been patented, and following an extensive development program, now has a proven record of performance in two world-scale nitration plants.

DESCRIPTION OF THE JET-IMPINGEMENT REACTOR

The general features of the jet-impingement reactor are shown schematically in Figure 1, and are described more completely elsewhere [4]. After introducing the reactants in a suitable way, the combined flows are commingled and made to flow through a series of baffles having inlet and outlet ports. The mixed flow forms into a series of jets, which are in turn directed against another baffle, or in some cases against each other.

The overall resultant flow field contains regions of high-intensity shearing action caused by the impingement or interplay of the jets. The shear rates are much higher than typically found in conventional stirred tank or static mixer-type reactors. In turn, the arrangement provides for a constant supply of fresh interfacial area between the reacting components, thereby enhancing the reaction rate and overall conversion efficiency of the reactor.

Another feature easily recognised is that the arrangement behaves as a plug flow reactor, with additional process benefits compared to back mix reactors [5].

The jet-impingement principle evolved from tests on scale-model prototype reactors, but in the main, development has been carried out in full-scale plants, with progress being made through stage-wise modification, dedicated plant trials, sample analysis and data interpretation, followed by further design modifications. This learning process was used in the subsequent construction of a world scale grass roots facility, where operating performance has exceeded expectation in every aspect.

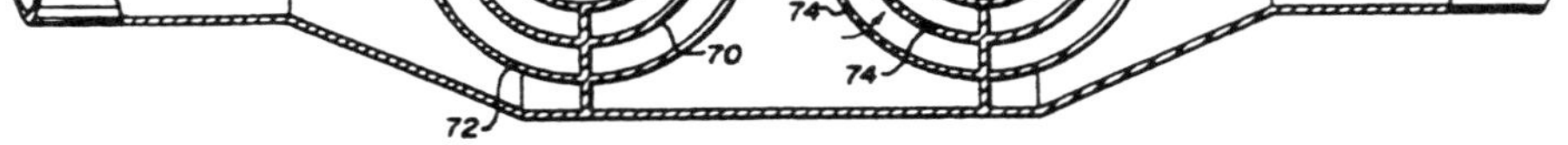

Figure 1. Schematic diagram of the jet-impingement reactor [4].

APPLICATION TO NITRATION PROCESSES

Nitration processes, involving heterogeneous liquid/liquid reactions between nitric acid and an immiscible organic compound, have been the first commercial application of the jet-impingement reactor principle. A schematic diagram of the process for production of MNB is shown in Figure 2.

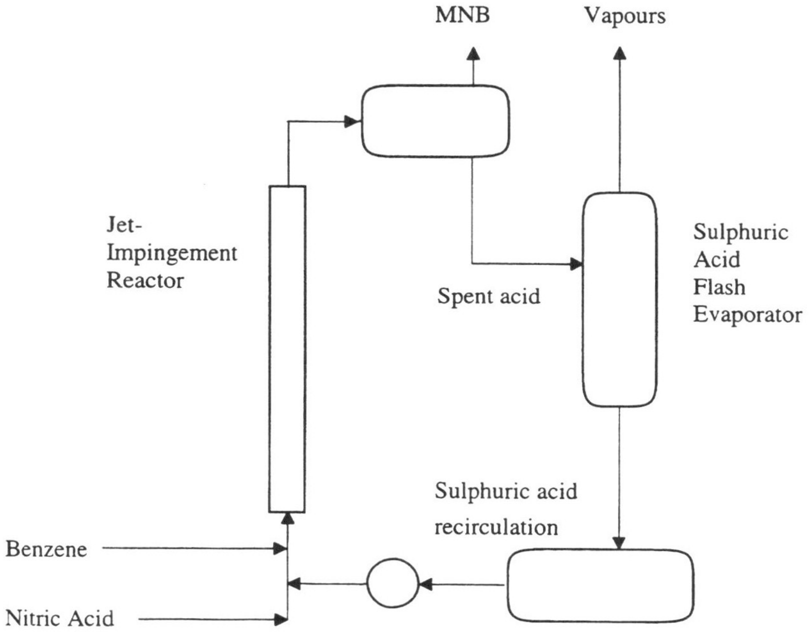

Figure 2. Application of the jet-impingement reactor to nitration reactions.

In this process, a mixture of benzene, nitric and sulphuric acid is circulated through the jet-impingement reactor, a MNB-acid decanter, and a sulphuric acid flash evaporator. The internal reactor elements produce a fine emulsion of benzene and MNB in the continuous acid phase, and are spaced and sized to achieve optimum conditions for mass transfer and process yield. Along with a synergistic combination of specific process conditions, described more completely in the patent literature [6], nitration rates at least an order of magnitude faster than those reported in the prior art technologies [7,8] have been demonstrated. Additional benefits achieved are; reduced equipment size and material inventory, lower operating and maintenance costs, increased process safety, and substantial reduction in by-product formation. More information on the process, and integration of the jet-impingement reactors is given in [5].

CONCLUDING REMARKS

The benefits of the jet-impingement reactor principle have been successfully proven at the plant scale, with an enviable record of on-stream performance. Refinement of reactor internal design has allowed optimal performance in one particular nitration process. Work is currently underway on developing the jet-impingement reactor for nitration of other aromatic hydrocarbons.

Other liquid/liquid and gas/liquid reactions such as alkylation or sulfonation should also potentially benefit from the high-intensity shear regions produced by the novel concept of jet-impingement.

REFERENCES

1. Giles, J.; Hanson, H.; Ismail, H.A.M., *Industrial and Laboratory Nitrations*; ACS Symposium Series; American Chemical Society; Washington, D.C. 1976, pp.190-209
2. Doraiswamy, L.K.; Sharma, M.M., *Heterogeneous Reactions, Vol.2*; John Wiley & Sons, New York, 1990
3. Narsimhan, G.; Nejfelt, G.; Ramkrishna, D.; *AICHE J.*; 1984, 30, pp.319
4. U.S. Patent 4,994,242; 1991
5. Guenkel, A.A.; Maloney, T.W.; *Nitration Symposium* ; ACS Symposium Series; American Chemical Society; Washington, D.C.; in press
6. U.S. Patent 5,313,009; 1994
7. U.S. Patent 4,973,770; 1990
8. U.S. Patent 4,091,042; 1978